C000242327

Molecules With Silly or Unusual Names

Molecules With Silly or Unusual Names

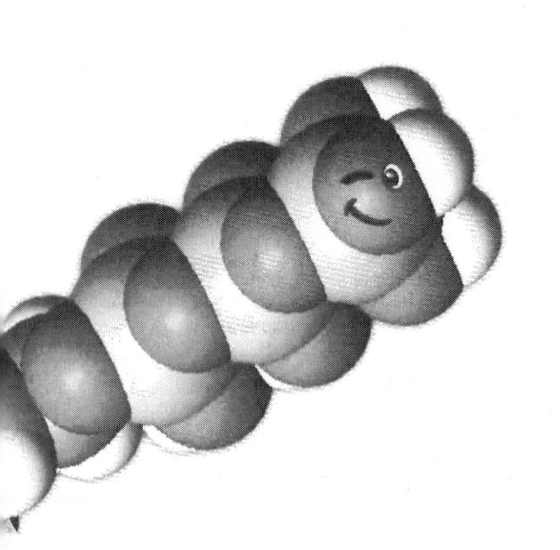

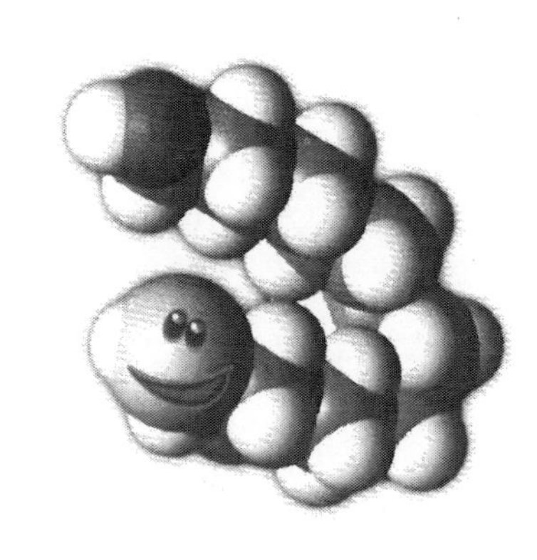

Paul W May

University of Bristol, UK

Imperial College Press

ICP

Published by

Imperial College Press
57 Shelton Street
Covent Garden
London WC2H 9HE

Distributed by

World Scientific Publishing Co. Pte. Ltd.
5 Toh Tuck Link, Singapore 596224
USA office: 27 Warren Street, Suite 401-402, Hackensack, NJ 07601
UK office: 57 Shelton Street, Covent Garden, London WC2H 9HE

British Library Cataloguing-in-Publication Data
A catalogue record for this book is available from the British Library.

MOLECULES WITH SILLY OR UNUSUAL NAMES

Copyright © 2008 by Imperial College Press
All rights reserved. This book, or parts thereof, may not be reproduced in any form or by any means, electronic or mechanical, including photocopying, recording or any information storage and retrieval system now known or to be invented, without written permission from the Publisher.

For photocopying of material in this volume, please pay a copying fee through the Copyright Clearance Center, Inc., 222 Rosewood Drive, Danvers, MA 01923, USA. In this case permission to photocopy is not required from the publisher.

ISBN-13 978-1-84816-207-5 (pbk)
ISBN-10 1-84816-207-3 (pbk)

Printed in Singapore by Mainland Press Pte Ltd

Contents

Preface vii

Acknowledgements ix

1. Molecules with Silly or Unusual Names 1

2. Minerals with Silly or Unusual Names 111

3. Proteins and Enzymes with Silly or Unusual Names 131

4. Genes with Silly or Unusual Names 143

5. Silly Molecules Names - the Game 175

Preface

Believe it or not, some chemists do have a sense of humour, and this book is a testament to that. On the following pages I'll show you some *real* molecules that have unusual, ridiculous or downright silly names. Some of the molecule names are bizarre, some make you wonder how supposedly serious scientists could possibly have called a molecule *that*, and some are very rude indeed!

Although the object of this book is mainly entertainment, I've tried to include a fair bit of science along the way, along with the references if you want to check that I didn't make it all up! Most of the molecules have their full structure shown, and with some I've included brief facts about their history and what they're used for. As well as molecules, I've also included some of my favourites from the vast number of minerals, proteins and genes that have often accidentally (or deliberately!) been given ludicrous names. Despite the humorous and often irreverent nature of the content, I'm hoping that the reader might unknowingly pick up quite a bit of chemistry along the way - learning chemistry 'by stealth', as it were.

So how did this book come about? Well, as you might have guessed, it started in the pub. I was having a drink with a few Chemistry colleagues one Friday after work in 1997, and the subject of the conversation somehow got around to molecules with unusual names. I mentioned that I'd just read about a mineral called *cummingtonite*, and everyone thought this was hilarious. A friend then bettered this by telling me that there was a genuine molecule called *arsole*. Within half an hour, between us, we'd come up with about 15 genuine molecules with ridiculous names. Back then, I'd just started writing webpages, so it was suggested that I should make a fun webpage of these, to share the joke with other friends who hadn't been fortunate enough to be at the pub that night. I did this, and within days I had people emailing me from all over the world saying they loved the site, and suggesting new molecules to add to the collection. Over the past decade the website expanded to nearly 250 molecules. The site gained infamy when it was featured in the science/humour sections of many top newspapers, including *The Times, The Guardian, USA Today*, as well as serious science magazines such as *Science, Scientific American* and *New Scientist*. *Playboy* magazine even paid me to reproduce some of the more dubiously named molecules in their humour column!

But how do the silly molecule names arise? Some names legitimately derive from the molecular structure, or from the location where the molecules were first discovered. Many come from the (Latin) name for the plant or animal species from which the molecules were isolated, or even from the name of the discoverer. Some molecules are given intentionally trivial names based on their structure, or simply as a result of the whimsy of chemists. Since trivial names predate formal naming conventions, they can be ambiguous or carry different meanings in different industries, geographic regions and languages. And the result of the (un)fortunate juxtaposition of innocent compound phrases can give rise to unintended but hilarious names (*e.g.* cummington + ite).

However, not everyone has access to the internet (yet), so I thought it might be a good idea to make a hardcopy version of the website to entertain (and educate) those people too. The original website is still going strong, and you are welcome to visit it:

http://www.chm.bris.ac.uk/sillymolecules/sillymols.htm

There, you'll find more information on the molecules, plus many colour diagrams and photographs, and often 3D animated molecular structures.

Enjoy.

Paul W. May (2008)

Acknowledgements

Although I used the web, plus many reference books and scientific journals to compile this book, there are a few references which require special mention.

- For a more detailed look at the naming of organic molecules, including expanded versions of some of the anecdotes described here, please see the excellent book: *Organic Chemistry: The Name Game*, by Alex Nickon and Ernest Silversmith (Pergamon, New York, 1987).
- *Elsevier's Dictionary of Chemoetymology: The Whies and Whences of Chemical Nomenclature and Terminology* by Alexander Senning (Elsevier, Amsterdam, 2007) is an invaluable source of information regarding the origin of molecule and mineral names.
- For minerals, the three websites which had almost all the information I could possible need are: www.mindat.org and www.webmineral.com and the *Virtual Museum of Minerals and Molecules* (www.soils.wisc.edu/~barak/virtual_museum).
- For proteins, the best references I found are the two online databases: *Entrez Gene*: (www.ncbi.nlm.nih.gov/sites/entrez) and the comprehensive *Online Mendelian Inheritance in Man* (www.ncbi.nlm.nih.gov/sites/entrez?db=omim), and for protein structures the huge collection at the *RSCB PDB database* (www.pdb.org) proved invaluable.
- For gene names there's *Entrez Gene* (see above), *Flybase* (flybase.bio.indiana.edu), *Flynome* (www.flynome.com), the *Clever Gene Names* website (http://tinman.vetmed.helsinki.fi/eng/intro.html),

and *The Interactive Fly Project* webpage
(www.sdbonline.org/fly/aimain/1aahome.htm).

Some of the images in this book were obtained from the Wikimedia Commons image library: (commons.wikimedia.org/wiki/Main_Page) and have been reproduced here under various licenses. See the appropriate web link to view the license and to obtain information about reusing these images.

- The Creative Commons Attribution Share-alike license: (creativecommons.org/licenses/by-sa/3.0/) and earlier versions.
- The GNU Free Documentation License (http://en.wikipedia.org/wiki/GNU_Free_Documentation_License)
- Public domain (http://en.wikipedia.org/wiki/Public_domain), because the author donated them for free use, the copyright had expired (normally ~70-100 years after the author/artist's death), or because they were taken/made by an employee of the US government during the course of the person's official duties.

Other images were purchased from Getty Images (www.gettyimages.com) and used on a Rights Managed basis, or are part (or modified versions) of the free Microsoft clipart collection.

I am grateful to the following people who emailed me over the past 10 years with ideas, suggestions or information about various silly molecule names:

Neil Brookes, Nicholas Welham, Andy Shipway, Lloyd Evans, Peter Sims, Mikael Johansson, Patrick Wallace, A. Haymet, Charles Turner, Tom Hawkins, Matthew J. Dowd, Chris Fellows, Bill Longman, Chris Valentine, Ben Stern, Beveridge, Justin E. Rigden, John Burgess, Neil Tristram, Michael Geyer, Matthew Latto, Neil Edwards, Han Wermaat and the Dutch Chemistry magazine 'Chemisch2weekblad', David Brady, Darren Sydenham, Bob Brady, Stephen O'Hanlon, Anthony Davis, Gerard J. Kleywegt, Van King, Melita Rowley, Bill Longman, Alan Howard Martin, Lionel Hill, Anthony Argyriou, Kristina Turner, Samuel Knight, Nicholas Welham, Victoria Barclay, Gavin Shear, Seranne Howis, Tom Simpson, Eric Walters, Allen Knutson, Prof Walter Maya, Michael F Aldersley, Carl Kemnitz, Michael J. Mealy, ShadowFox, Anton Sherwood, Andrew Reinders, A. Rich, Adrian Davis, Iain Fenton, John P Oliver, Peter Macinnis, Melissa Harrison, Eric Walters, Liz Parnell, Victor Sussman, Amy Roediger, Andrew P. Rodenhiser, Jesper Karlsson, Rene Angelo Macahig, Kay Brower, Martin A. Iglesias Arteaga, Mark Minton, Matti Lepisto, Stephen Yabut, Phil Van Es, Andrew P. Rodenhiser, Marcel Volker, Hazel Mottram, Phil van Ess, Alan Plante, Shawn McClements, (and his high school class), Andrew Walden, Florian Raab, Bo Ohlson, Robin Brown, Joerg Fruechtel, Nicholas J. Welham, Ronald Wysocki, Chris Scotton, Terry Frankcomb, Juan Murgich, Phillip Barak, John Lambert, Tanuki, 'Sparkly' Sally Ewen, Sean, Kay Dekker, Birgit Schulz, Lars Finsen, Steve Colley, Calli Arcale, Ogpusatan, Danny Sichel, Mark Baxter, Janet McBride, Ewart Shaw, Steve Stinson, John Figueras, Suds Mixer, Colin Metcalf, Jarrod Ward, Andy

Dicks, Amanda Musgrove, Greg Valure, Richard Williams, Geraldine VdA, David Bradley, Jeremy Bracegirdle, Mathias Disney, Christopher C. Wells, Gene Morselander, Tanuki the Raccoon-dog, Friedrich Menges, Steven A. Hardinger, Hans Hillewaert, Germanicus Hansa-Wilkinson, Christopher Putnam, Peter Traill, John Gosden, Ken Weakley, Thomas Vaid, Enrique Pandolfi, Indranil Sen, Blair Boehmer, Shefa Gordon, Leigh Arino de la Rubia, Stuart Kidd, Grant Little, Sean Pearce, Simon Cotton, Polyploid2, Martin Harris, Boaz Laadan, Victor Sussman, Brian Jackson, Terence Bartlett, Joseph Wiman, Nicholas J. Welham, Mark Johnston, Gary Randall, 'Plutonium' Page Sebring, Jonathan Montgomery, Isaiah Shavitt, Susanne Wikman, Mark Croker, Michael Bailey, Larry Baum, Alex Yuen, Neil Brew, Satan's Little Helper, Michael Stewart, Karel Vervaeke, Harmen Lelivelt, Fernando Perna, Andy Cal, J.J. Keating, Lee Flippin, Joris van den Heuvel, Geoff Hallas, Andrew Byro, Eric van der Horst, Matthew Piggott, Stephen F.T.M. Thompson, Allart Kok, Andrew Patterson, Jerry Van Cleeff, Victor Nikolaev, John L. Meisenheimer, Sr., Richard Cammack, Han Lim , Tom, John Moody, Erik Holtzapple, Jim Gobert, Tim Lyon, Marc Kaminski, Peter Lykos, David Ball, Sophie Weiss, Mackay Steffensen, Kyle Daly, Peter Keller, Lily Zhou, Hope Nesmith, Kenneth Koon, Matt Wright, Michael F Aldersley, Fox, James Waterman, Thomas Jeanmaire, Marc Schaefer, Ken Ruiz, Kutti, Antony Bigot, Dave Chapman, Jason Stouffer, Warut Roonguthai, Willem Schipper, Geoff Hallas, Wendy Hunt, Leigh S. Arino de la Rubia, Martin Lee, Chris Miller, Rob Saunders, Ian, David French, Vincent Schüler, Daniel Manke, Neil Anderson, David Bromage, Debbie Radtke, Keith Bromley, Iain Fielden, Bill Edmonds, Tavi, Mark Isaac, Anthony Nicholls, Joe Fortey, Duncan Wiles, Helen Webb, Shane Liddelow, Arjen van Doorn, Brandi Baros, Stephen Ashworth, Eric Kaufman, Ian Livingstone, Jenni Vedenoja, Joe Bobich, Thomas Schneider, Mirela Matecic, Anne Gorden, Vincent Schüler and Peter Rice.

If you liked this book, you might also like to check out:

- The original *Molecules with Silly nd Unusual Names* website (http://www.chm.bris.ac.uk/sillymolecules/sillymols.htm)
- *Curiosities of Biological Nomenclature* (home.earthlink.net/~misaak/taxonomy.html)
- *A Barrel Full of Names* (www.jimwegryn.com/Names/Names.htm)
- *Funny or Curious Zoological Names* (cache.ucr.edu/~heraty/menke.html)
- *Doug Yaneda's Biological names page* (http://cache.ucr.edu/~heraty/yanega.html#"LIGHTER"%20LINKS)

1. Molecules with Silly or Unusual Names

Arsole

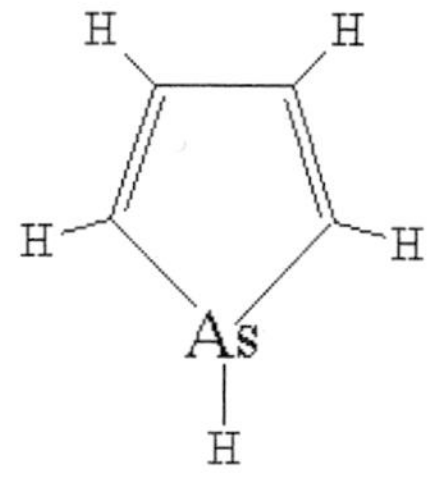

Yes, believe it or not, there is actually a molecule called *Arsole*... and it's a ring! It is the arsenic equivalent of *pyrrole*, and although it is rarely found in its pure form, it is occasionally seen as a sidegroup in the form of organic *arsolyls*. For more information, see the paper with probably the best title of any scientific publication I've ever come across: "Studies on the Chemistry of the Arsoles"[1].
Contrary to popular belief, new research[2] shows that *arsoles* are only moderately aromatic... Incidentally US patent number US 3 412 119 by the Dow Chemical Company is entitled "Substituted Stannoles, Phospholes, Arsoles, and Stiboles" - I didn't know there was a substitute for an *arsole*... Furthermore, the structure where *arsole* is fused to a *benzene* ring is called *benzarsole*, and apparently when it's fused to 6 *benzenes* it would be called *sexibenzarsole* (although that molecule hasn't been synthesised yet). Two other poisonous arsenic molecules include the simple hydrides, called *arsine* (AsH_3) and *arsorane* (AsH_5).
And on a related theme, there's an *aryl selenide* compound with the superb shorthand of *ArSe*, which is both toxic and smelly. The paper it comes from[3] was published by authors from, of course, the University of Aarhus!
Also, the related molecule, *phosphole* (which just replaces As with P) is quite amusing if you are a French speaker, since it's pronounced the same as *fausse folle*, where *fausse* means 'false' or fake' and *folle* means both a 'crazy woman' and a 'drag-queen' or 'ladyboy'.

Adamantane

Adam Ant in the 1980s
(Copyright Getty Images, used with permission Photo: Chris Walter) .

This molecule always brings a smile to the lips of chemistry students when they first hear its name, especially in the UK. For those not in the know, ***Adam Ant*** was an English pop star in the early 1980's famous for silly songs and strange make-up.

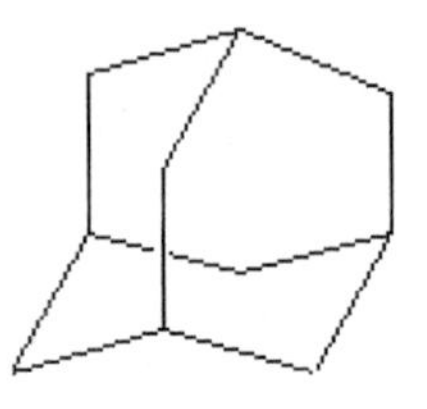

Two *adamantanes* fused together make *diamantane,* which was originally known as *congressane* since its synthesis was announced at an IUPAC congress in the 1960s. the name was changed since there are lots of congresses, and it seemed a bit silly to name this chemical after just one of them - or maybe IUPAC looked up the word 'congress' in the dictionary [4] and were worried about its alternative meaning...

Bastardane

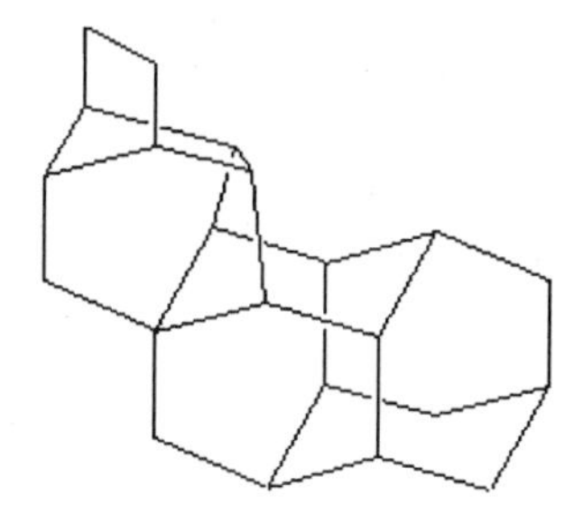

This is actually a close relative of *adamantane*, and its proper name is *ethano-bridged noradamantane*. However, because it had the unusual *ethano* bridge, and was therefore a variation from the standard types of *diamondoid* hydrocarbon cage structures, it came to be known as *bastardane* - the "unwanted child" [5].

In fact, the paper where it was first reported was entitled "Nonacyclodocosane, a Bastard Tetramantane" [6].

Megaphone

Is this perhaps the loudest molecule of all? Despite having a ridiculous name, the molecule is quite ordinary. It gets its name from being both a constituent of *Aniba Megaphylla* roots and a ketone[7].

Unununium

I know this is technically an element, not a molecule, but it had such a ridiculous name I thought I'd include it. This is actually element number 111, and was called *unununium* by the IUPAC temporary systematic name[8] before it was recently renamed *roentgenium*. This is a pity, because if it formed ring or cage structures, previously we might have ended up with *unununium* onions...

Arabitol

(Rabbit photos copyright Getty Images, used with permission. Photographer Jo Sax).

No, this has nothing to do with rabbits - it's an organic alcohol that's one constituent of wine. It's also known as *pentahydric alcohol*. A related sugar molecule, *arabinose*, also has nothing to do with rabbits, nor with the size of a Rabbi's nose (A Rabbi Nose).

Moronic Acid

This is a *triterpenoid* organic acid that is found in *Pistacia* resin, and is therefore of interest to people studying archaeological relics, shipwrecks and the contents of ancient Egyptian jars.[9] I'm not sure how it gets its stupid name, but maybe it was first extracted from Mulberry trees (*Morus*). Derivatives of this are called *moronates*, as in 'which moron-ate the contents of this jar?'

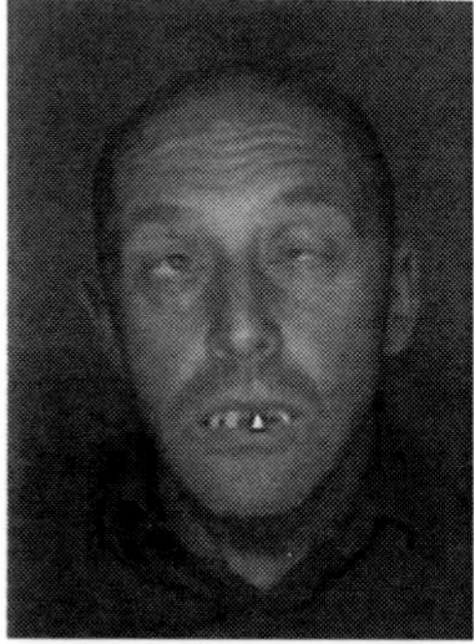

Maybe he's been eating too much moronic acid...
(Copyright Getty Images, used with permission.
Photo: Christopher Thomas).

Erotic Acid

Unfortunately, this isn't the world's best aphrodisiac. Its correct name is *orotic acid*, but it has been misspelled so often in the chemical literature that it is also known as *erotic acid*! Another name for it is *vitamin B13*. Apparently, if you add another carbon to it, it becomes *homo-erotic acid*...

Munchnones

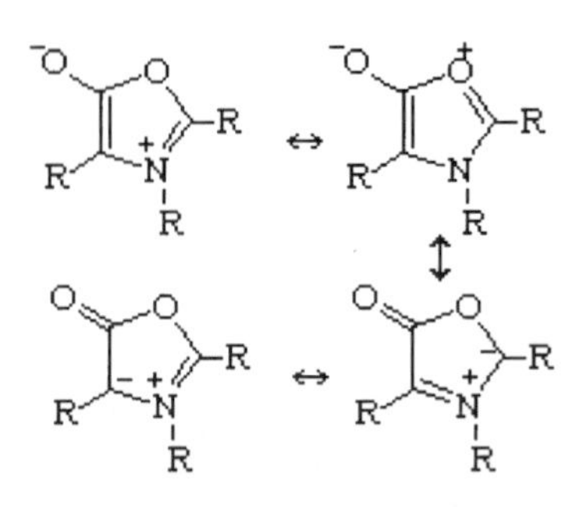

No, these aren't the favourite compound of the *Munchkins* from *The Wizard of Oz*, but are in fact a type of *mesoionic* compound. These are ring structures in which the positive and negative charges are delocalised, and which cannot be represented satisfactorily by any one polar structure. They got their name when the chemist[10] who first synthesized them called them after the city Munich (München), after similar compounds were called *sydnones* after Sydney.

Munchkins present Judy Garland with a lollipop (and not a *munchnone*) in the film 'The Wizard of Oz'.
(Copyright Getty Images, used with permission).

Fucitol

Although this sounds like what an undergraduate chemist might exclaim when their synthesis goes wrong, it's actually an alcohol, whose other names are *L-fuc-ol* or 1-*deoxy-D-galactitol*. It gets its wonderful trivial name from the fact that it is derived from the sugar *fucose*, which comes from a seaweed found in the North Atlantic called 'Bladderwrack' whose Latin name is *Fucus vesiculosis*.

Buckminsterfullerene

This is the famous soccerball-shaped molecule that won its discoverers' the Nobel prize for Chemistry in 1996. It is named after the architect Buckminster Fuller who designed the geodesic dome exhibited at Expo '67 in Montreal. It was from this building that Sir Harry Kroto got the idea how 60 carbon atoms could be arranged in a perfectly symmetrical fashion. Because the name of the molecule is a bit of a mouthful, it is often referred to just as a *Bucky Ball.*

Buckminster Fuller's geodesic dome at Expo '67 in Montreal.
(Copyright Getty Images, used with permission. Photo: Michael Rougier).

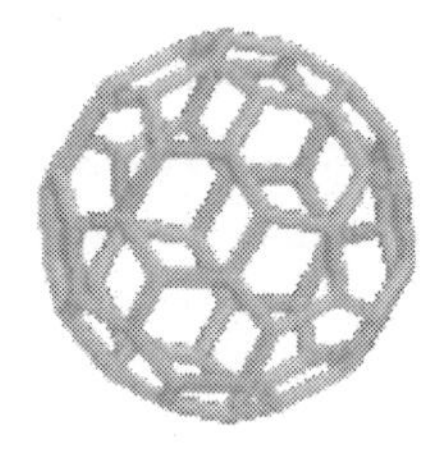

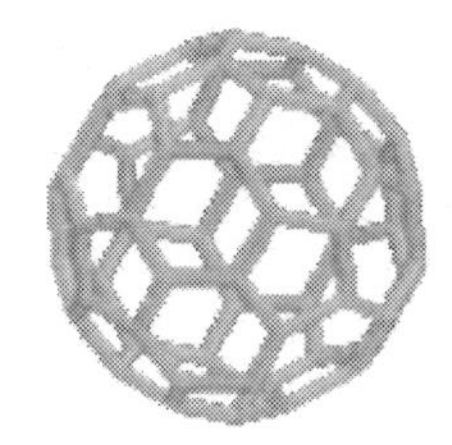

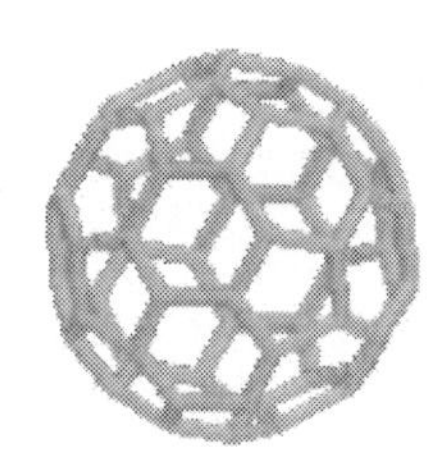

It's also known as *footballene* by some researchers [11]. In fact, there is now a whole '*fullerene zoo*', with oddly coined names[12], including: *buckybabies* (C_{32}, C_{44}, C_{50}, C_{58}), *rugby ball* (C_{70}), *giant fullerenes* (C_{240}, C_{540}, C_{960}), *Russian egg* or *bucky onions* (balls within balls), *fuzzyball* ($C_{60}H_{60}$), *bunnyball* ($C_{60}(OsO_4)$(4-*t-butylpyridine*)$_2$), *platinum-burr ball* ($\{[(C_2H_5)_3P]_2Pt\}_6C_{60}$) and *hetero-fullerenes* (in which some carbons are replaced by other atoms).

There is also a *fullerene* paper in which the authors describe a method for severing two adjacent bonds in C_{60}, entitled "There Is a Hole in My Bucky".[13]

Putrescine, Cadaverine, Spermine & Spermidine

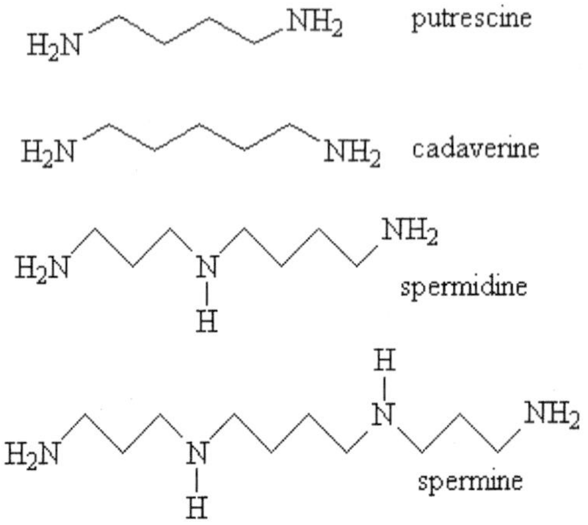

H_2N NH$_2$ putrescine

H_2N NH$_2$ cadaverine

H_2N N(H) NH$_2$ spermidine

H_2N N(H) N(H) NH$_2$ spermine

Putrescine originates in putrefying and rotting flesh, and is quite literally, the smell of death. It is one of the breakdown products of some of the *amino-acids* found in animals, including humans. Although *putrescine* is a poisonous solid, as flesh decays its vapour pressure becomes sufficiently large to allow its disgusting odour to be detected. It is usually accompanied by a poisonous syrupy liquid with an equally disgusting smell called *cadaverine* (named after the cadavers that give rise to it). Surprisingly, both these molecules of death also contribute towards the smells of some living processes. Since they are both poisonous, the body normally excretes them in whatever way is quickest and most convenient. For example, the odour of bad breath and urine are 'enhanced' by the presence of these molecules, as is the 'fishy' smell of the discharge from the female medical condition *bacterial vaginosis*. *Putrescine* and *cadaverine* also contribute to the distinctive smell of semen, which also contains the related molecules *spermine* and *spermidine*. So sex and death are indeed, closely related.

Curious Chloride and Titanic Chloride

The trivial name for some curium compounds can be either *curous* or *curious*, so *curium trichloride* becomes *curious chloride*. However the only curious property it has is that it's sufficiently radioactive that a solution, if concentrated enough, will boil spontaneously after a while. (I wonder if a molecule with 2 Cm atoms in would be called *bi-curious*...?)

In a similar way, titanium compounds can be *titanic*, so we get the wonderfully named *titanic chloride*, $TiCl_4$. In the titanium industry, $TiCl_4$ is often known as 'tickle'. Furthermore, *curium oxides* are called *curates*, so the titanium compound would be *titanic curate* (a huge vicar?), and since curium can have more than one valency we could end up with *curious curates*. But I'm sure these are already a well-known phenomenon...

In a similar way, some nickel compounds can be referred to as *nickelous* - so we get compounds like *nickelous sulfate* (a nice guy by all accounts...).

Traumatic acid

This is a plant hormone which causes injured cells to divide and help repair the trauma - hence its name, and its synonym 'wound hormone'.

Gossypol

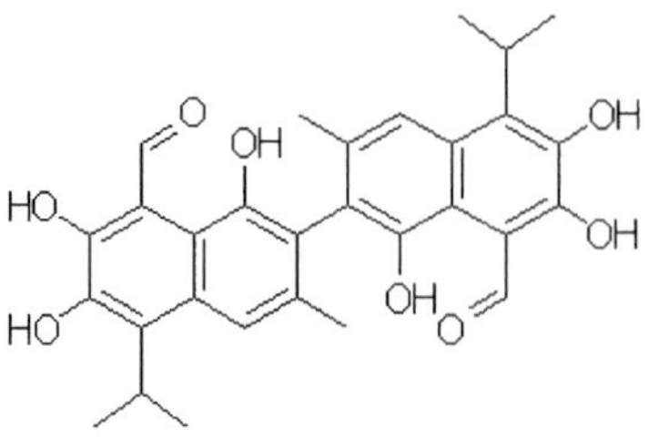

This ridiculously named molecule is found in cotton seeds. It was used as a male contraceptive in China, but was never used in the West (and may have since been banned in China as well), since its effects were permanent in about 20% of patients! Its name originated from being present in the flowers of the Indian cotton plant *Gossypium herbaceum L.*

Apart from its contraceptive effects, *gossypol* has properties that might make it useful in treating a number of ailments, including cancer, HIV, malaria and some bacterial/viral illnesses. Related to this molecule are the equally strangely named *gossypetin* and *gossypin*. I always thought *gossypin* was frowned upon in polite labs....

Bastadin-5

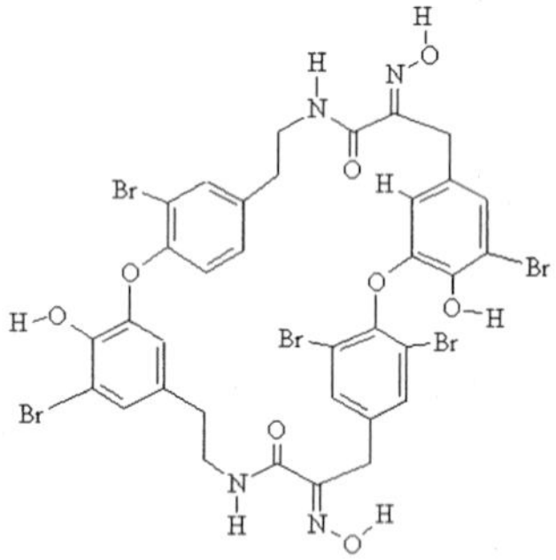

This is just one of a number of *bastadins*, which are molecules isolated from the marine sponge *Ianthella basta*[14]. They possess antibacterial, cytotoxic and anti-inflammatory properties.

Vomicine

This poisonous molecule gets its name from the nut *Nux Vomica,* which is the seed of a tree found on the coasts of the East Indies. The seeds are sometimes called 'Quaker buttons', and are a source of *strychnine* as well as the emetic *vomicine.*

Skatole

This molecule's name comes from the Greek for manure (*skor*). Its proper name is 3-*methylindole,* but it gets its trivial name from the fact that it is a component of feces. Surprisingly, it is also found in coal tar and beetroot (!), and can be obtained synthetically by mixing egg albumin and KOH. As you might guess, *skatole* consists of white to brownish scales which are soluble in hot water. Apparently, coriander can be used to cover up bad smells such as these, as testified in the classic paper by Kohara *et al*: "Deodorizing Effect of Coriander on the Offensive Odor of the Porcine Large Intestine." [15]

Rhamnose and Rhamnetin

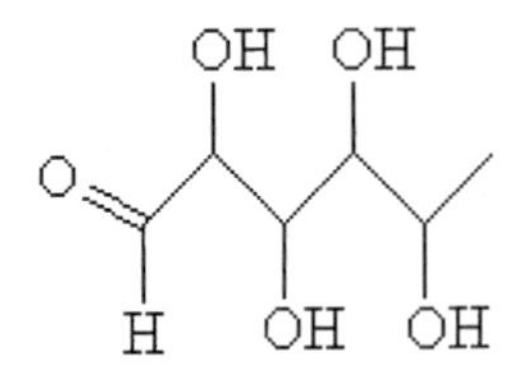

Rhamnose sounds like the molecule that's created when you walk into doors...in fact it's a type of sugar. It's made by hydrolysis of a *glucoside* found in buckthorn berries. The Greek for buckthorn berry is *rhamn*, and the *-ose* is because it's a sugar.

From the same berry (*Rhamnus cathartica*) comes *rhamnetin*. This molecule with an amusingly double-entendre name (ram'n it in) is a yellow pigment used in the dye industry.

Sexithiophene

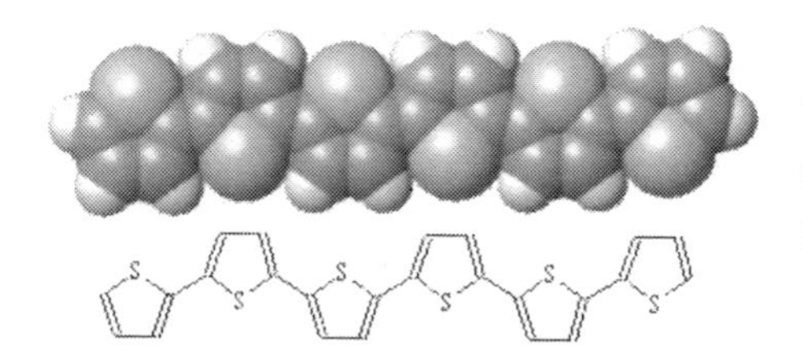

This is a 'sexi' molecule - which means it has 6 sub-units, in this case of *thiophene* rings.

Because of its conjugated system of double bonds, this organic molecule conducts electricity quite well. As a result, it is one of a number of similar molecules being studied for possible uses in organic polymer electronics. Incidentally, the Latin for 5 sub-units is *quinque* (pronounced 'kinky'), so by adding one subunit a *quinque* molecule becomes *sexi*... Nine units would be called *nonakis-*, which shows what always happens if you try to take things too far!

Bis(pinacolato)diboron

A real *pina colada*
(Copyright Getty Images, used with permission. Photo: Andy Caulfield).

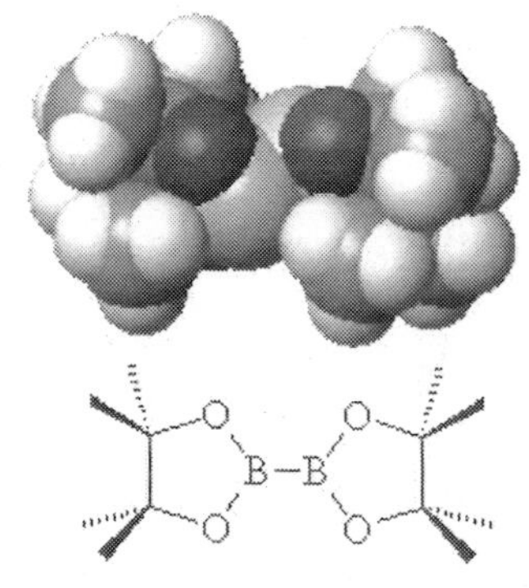

Although it sounds like it, this isn't the active ingredient in a *pina colada* cocktail. Rather, it is a versatile reagent for the preparation of *boronic esters* from *halides*, the diboration of *olefins*, and solid-phase Suzuki coupling [16]. The name comes about since it's a derivative of the molecule *pinacol*. Incidentally, a proper *pina colada* cocktail is a concoction of pineapple juice, coconut milk and rum, often served with crushed ice and a little paper umbrella stuck in the glass.

Crapinon

Crapinon (also known as *Sanzen*) is another molecule with an excellent name, and is apparently used therapeutically as an anticholinergic. These are drugs which dry secretions, increase heart rate, and decrease lung constriction. More importantly, they also constipate quite strongly - so 'crappy-non' is most appropriate.

It would be nice to think that this molecule could find an alternative use as a toilet cleaner (as in "Who's been crapinon the seat?").

Sparassol

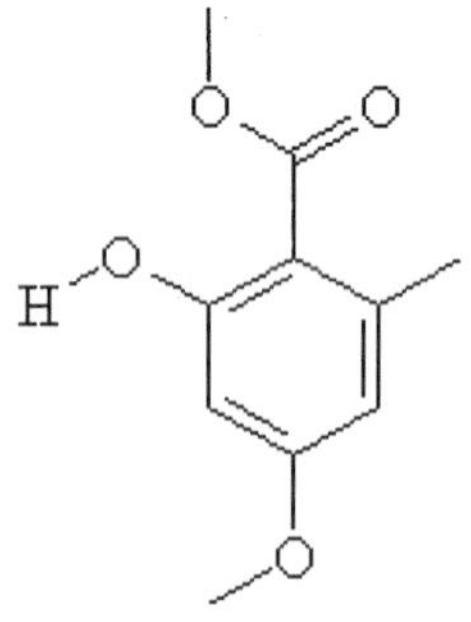

This molecule sounds like what you'd need the day after eating a very hot curry (a spare-assol). *Sparassol* is an antibiotic produced by the fungus *Sparassis ramose*, which gets its name from the Greek word *sparassein* meaning to tear or rend. Maybe this is the origin of the phrase 'to tear someone a *sparassol*'?

Phthalic Acid

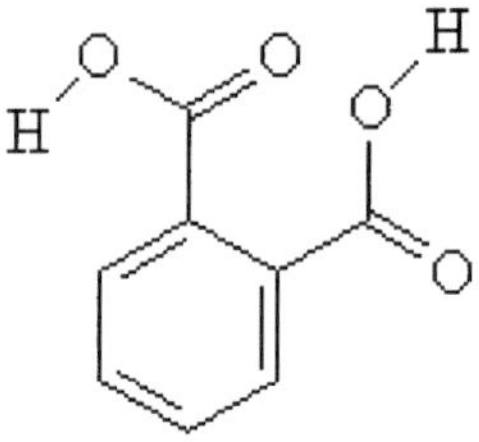

Phthalic acid

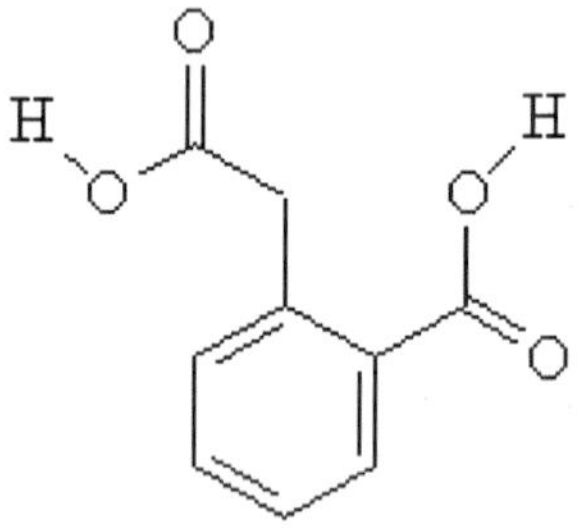

Homo-phthalic acid

This molecule is often pronounced with a silent 'th' for comic effect. I wonder if *phthalyl* side-groups have a shorthand symbol in chemical structures, in the same way that *phenyl* groups are shortened to -Ph? If so, would it be a *phthalic symbol...*? Again, adding an extra carbon makes *homo-phthalic acid* - say no more...

Periodic Acid

Ok, I know it should really be *per-iodic acid*, but without the hyphen it sounds like it only works some of the time... It has also been described as the acid that is extracted after boiling down old Periodic Tables found in chemistry lecture halls and laboratories.

Psicose

This molecule has nothing to do with Norman Bates, but is a sugar which gets its name because it's isolated from the antibiotic *psicofurania*. Its other name is *ribo-hexulose*.

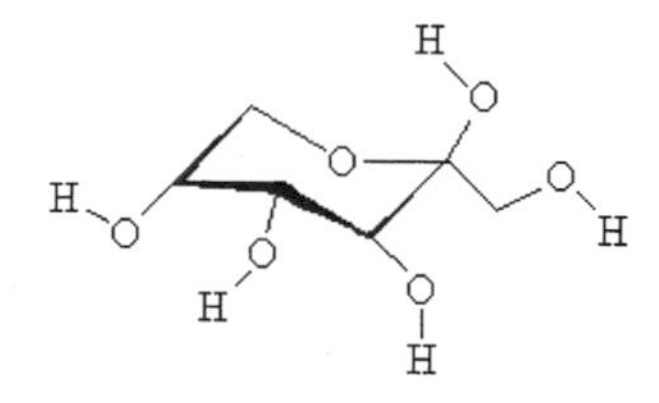

The Bates' Motel from the film 'Psycho'.
(Copyright Getty Images/Paramount Pictures, used with permission).

Commic Acid

This molecule's always good for a laugh! It gets its 'commical' name since it's a constituent of the plant *Commiphora pyracanthoides*, one of the Myrrh trees [17]. When reduced to the *aldehyde,* I presume the product would be named *commi<u>cal</u>*?

Gigantine

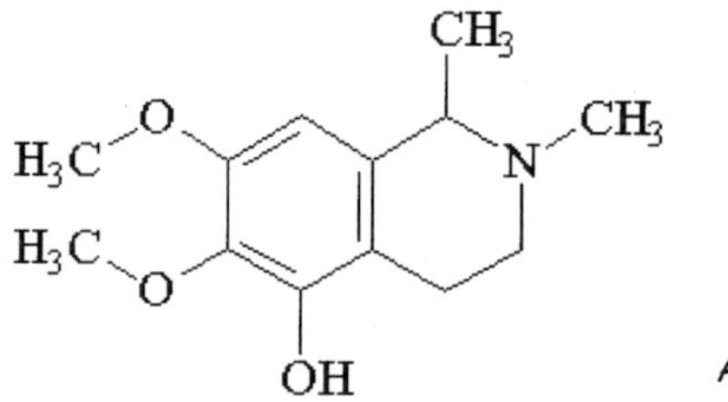

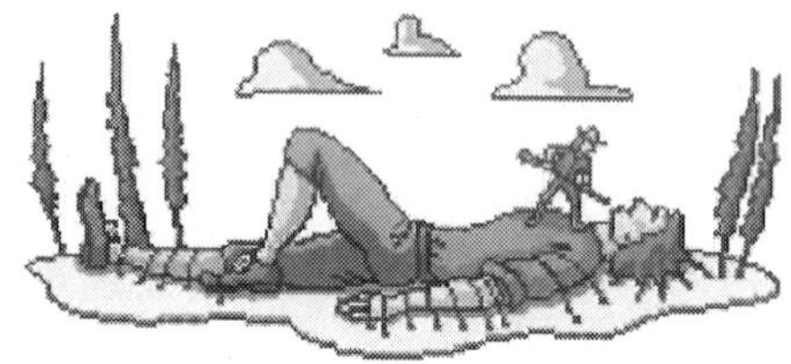

A Lilliputian needs some *gigantine* to keep up with Gulliver...

This chemical comes in very, very large bottles! It's from the cactus *Carnegia gigantea* [18], and is a hallucinogen - so perhaps it just makes everything appear gigantic.

Fruticolone

This sounds like what you get after a baked bean meal...but it actually gets its name from being both a constituent of the plant *Teucrium fruiticans* and a ketone [19]. There's also a variant of this called *isofruticolone.*

Nonanone

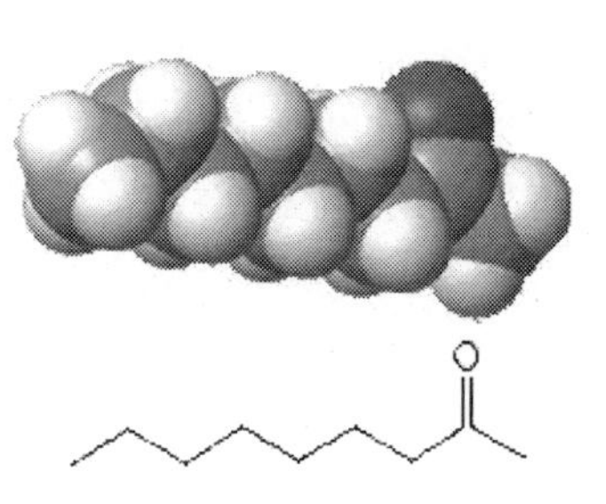

Although maybe not quite as silly as some of the other molecular names, I like this one for its n-n-nice alliteration. Many *nonanones* act as alarm pheremones in wasps, ants and bees.

Interestingly, in Danish, Norwegian, Swedish and German molecular names are spelt without the end "e" (*e.g. butane* is *butan, etc.*). Therefore, *nonanone* becomes '*nonanon*', and is quite an exceptional molecule name, being spelled the same way forwards and backwards - a palindromic molecule! The molecule shown is 2-*nonanone,* but 5-*nonanone* with the C=O group in the middle would be the same forwards as well as backwards, thus being palindromic in spelling and in structure!

Fukugetin

This chemical with a most amusing name is also called *morellofavone*, and is a constituent of the bark of the *Garcenia* species of tree [20]. Its glucoside goes by the equally wonderful name of *fukugiside*.

Nudic Acid

A molecule for naturists? This is an antibiotic derived from mushrooms, one of which (*Tricholomo nudus*) was the origin of its name [21].

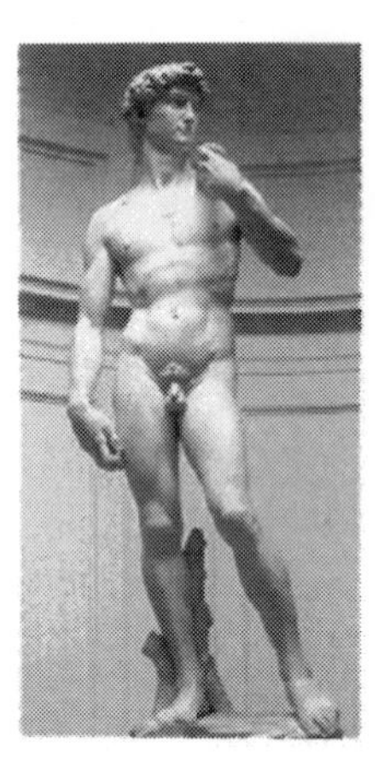

Right: Michelangelo's statue of David.
(Photo: David Gaya [22]).

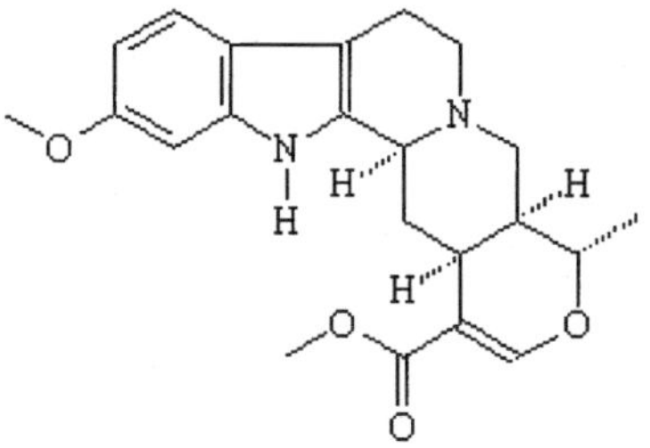

Pubescine

Also known as *reserpinine*, it got its name since it was extracted from the plant *Vinca pubescens*[23]. I don't know much else about *pubescine*, but I bet it forms short, curly crystals...

Funicone

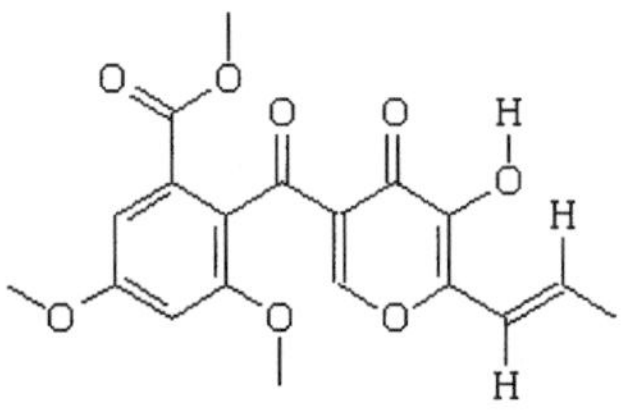

This gets its name, not from being funny and cone-shaped, but because it's the metabolism product of the fungus *Penicillium funiculosum*.

Powder of Algaroth

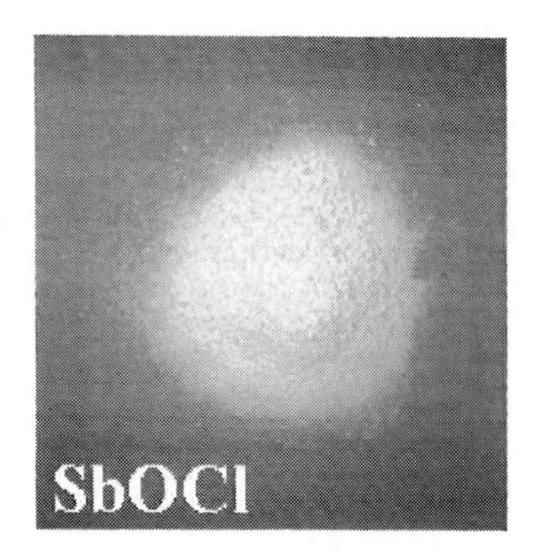

This sounds like one of Harry Potter's wizardly potions, and is the archaic name for *antimony oxychloride* (SbOCl). It was named after the Italian chemist Vittorio Algarotti who discovered its use as a medical emetic which purges violently both 'upwards' and 'downwards'... Perfect for a party prank, then...

Didymium

Nd + Pr

Pr + Nd

Nd + Pr

This is another old term, this time for a mixture of neodymium and praseodymium, which, due to their similar chemical properties were inseparable for many decades. The name comes from the Greek word for twin (*didymos*).

Melon

Melon is a flame-retardant chemical. Its name is coined by variation from that of a similar molecule called *melem*, which got its name from another similar molecule *melamine*, which got its name from *melam*, which was just an arbitrary name! Would an aqueous solution of *melon* be called a *water-melon*?

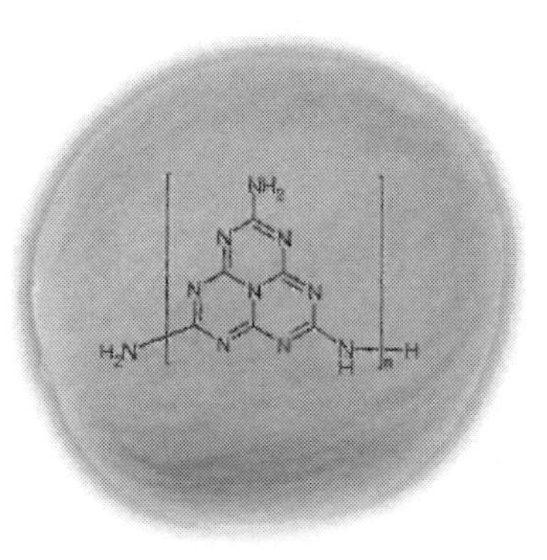

Melon drawn on a melon.
(Photo of melon: Knutux [24])

Hanuš Reagent

Dudes, this reagent is, like, totally heinous! It's, like, a totally equimolar mixture of *iodine* and *bromine* in *glacial acetic acid*, named after the most excellent, and definitely non-heinous Czech chemist Josef Hanuš. It's used to measure the unsaturation (number of double bonds) in organic substances. Radical !

Saponin

This soapy molecule gives rise to a classic chemists' greeting: "Hey, what's *saponin*?" There are many types of *saponins* found in various plants. They get their name from *sapon* meaning 'soap' in Latin, due to the fact they form a stable soapy froth when mixed with water. The one shown above is called 3-*O*-β-*glucopyranosyl arjunolic acid*.

Strawberry Aldehyde

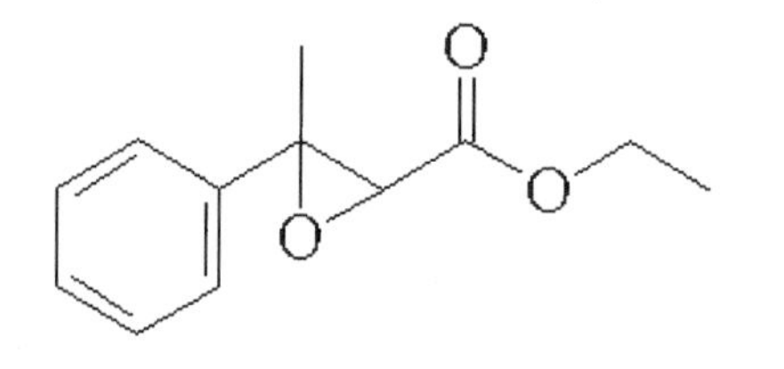

(Photo: FoeNyx [25])

This molecule's proper name is *ethyl methylphenylglycidate*, and is used in the food industry as artificial stawberry flavour. Despite its common name of *strawberry aldehyde*, it's not an *aldehyde*, being classified as an *ester* and *epoxide*, and it doesn't come from strawberries, although it does taste of them.

Spamol

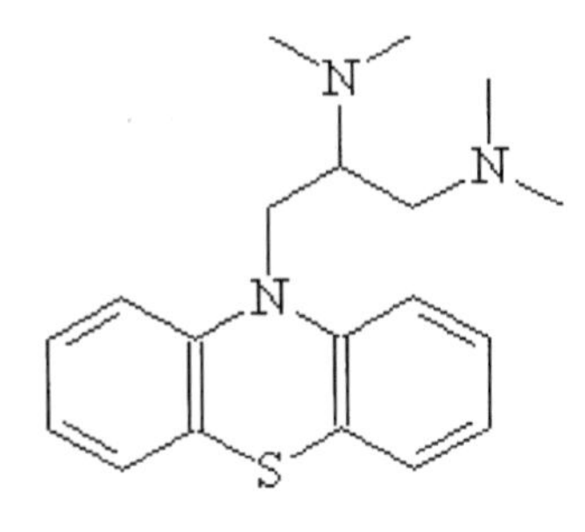

(Photo: PWM)

Could this be Monty Python's favourite molecule? *Spamol* might also conjure images of unwanted "Make Money Fast" emails circulating the globe ('spam-all'). Its other names are *aminopromazine, lispamol* or *lorusil,* and it's actually used as an anti-spasmodic therapeutic agent (or should that be anti-*spams*odic...?).

Fukiic Acid

Fuki is the Japanese word for the butterbur flower, and *Fukiic acid* is the hydrolysis product from this plant, *Petasites japonicus.* Interestingly, further oxidation of this produces the wonderfully named *fukinolic acid.* (I wonder if *fukanolic* is anything like alcoholic...). Since the conjugate base of *fukinolic acid* is *fukinolate,* it's probably about time we stopped.

Housane, Churchane and Basketane

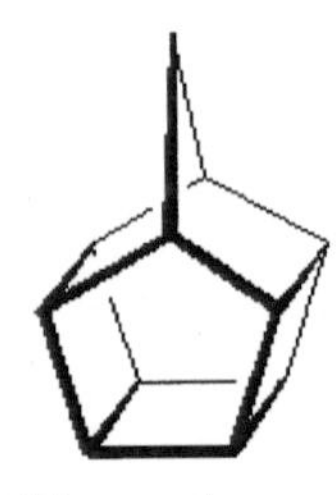

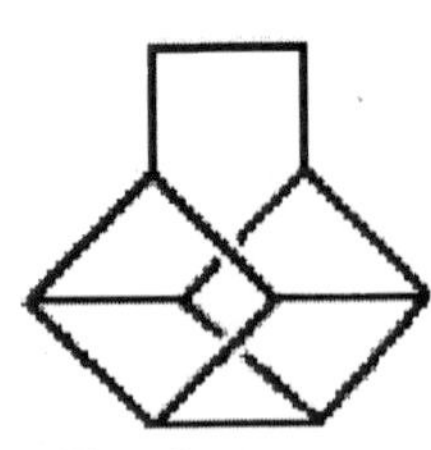

Obviously, these molecules get their name from their shapes[26]. Although I do think that *housane* (how sane?) should be closely linked to *psicose*, above.

Windowpane

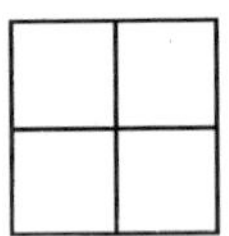
Windowpane

On the same theme, *windowpane*, C_9H_{12}, gets its name from its resemblance to a set of windows[27], but unfortunately the bond strain is so great that it has never been synthesised. It's more accurately known as *fenestrane* from the Latin *fenestra* (= window)[5]. But the version with a corner carbon missing, C_8H_{12}, has been made, and goes by the name *broken window*[28]. Interestingly, a hypothetical derivative of *windowpane* has been suggested which includes a double bond, and this would of course be called *windowlene*... (This mustn't be confused with the real molecule, *vindolene*, $C_{25}H_{32}N_2O_6$ which is an alkaloid derived from periwinkles).

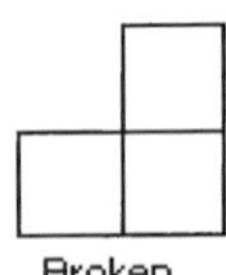
Broken Windowpane

Germane

Germane GeH_4

Right: The molecule

Below: Definition from the Pocket Oxford Dictionary

germane *adj.* (usu. foll. by *to*) relevant (to a subject).

This is a particularly *relevant* molecule that is *pertinent* and *has a bearing on* a number of inorganic reactions...

Godnose

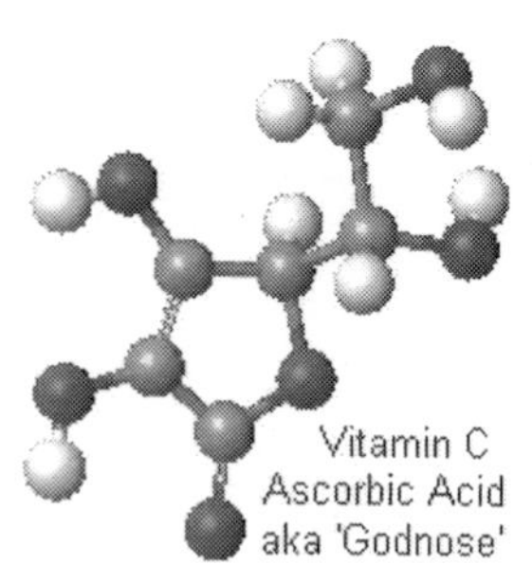

Vitamin C
Ascorbic Acid
aka 'Godnose'

Ok, so this is a bit of a cheat, since it's not an official molecular name...but it makes a nice story [5]. When Albert Szent-Gyorgyi isolated *ascorbic acid* and published his findings, he called the new substance '*ignose*' from '*ignosco*' meaning 'don't know' and '-ose' meaning a sugar.

When the journal editor refused to accept *ignose* as a sensible name, Szent-Gyorgyi suggested *godnose* instead! Alas, the editor was neither imaginative nor humorous, and suggested that a more proper name had to be used [29]. The structure of the carbohydrate was elucidated and the sensible name given to it was *hexuronic acid* (hex = six). During the same period (1928–1931), Charles Glen King of Columbia University isolated *vitamin C* from lemon juice, and soon realised that *hexuronic acid* and *vitamin C* were identical.

Incidentally, the name *ascorbic acid* derives from '*a-*' (against, or counteracting) and the Latin *scorbutus*, meaning the disease scurvy, since eating it prevents scurvy, as the old English sailors (the original 'Limeys') discovered. This word was itself derived from Old Norse (*skyr* = old curdled milk) and *bjugr* (edema) – referring to the ancient Viking belief that scurvy was caused by the sailors' diet of old curdled milk.

Diabolic Acid

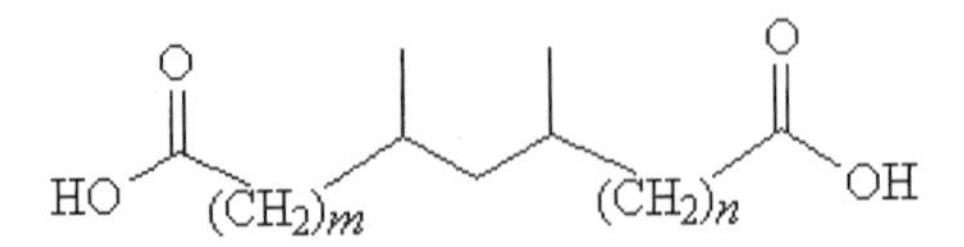

Diabolic acids are actually a class of *dicarboxylic acids* with the general structure shown above, where the *m* and *n* chains can have different lengths and may contain double bonds. They were named after the Greek *diabollo*, meaning to mislead, since they were devilishly difficult to isolate using standard gas chromatography techniques [30].

Domperidone

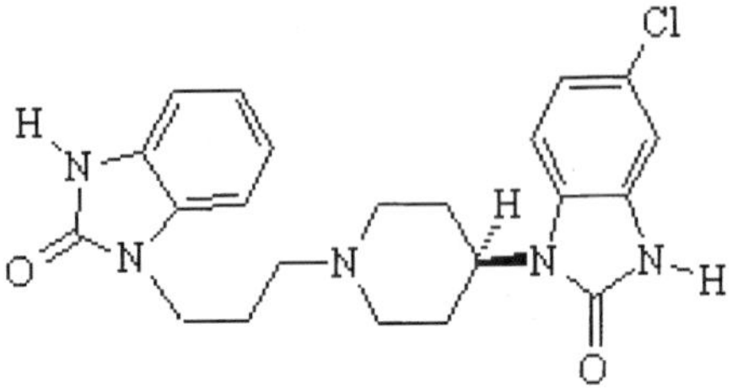

Left: Champagne bottles[31]

This molecule sounds like it should be the active ingredient in *Dom Perignon* champagne, but it's actually an anti-emetic drug. It is also used to promote the production of breast milk in lactating (or non lactating women), and even to induce lactation in a male!

Fluorene and Theobromine

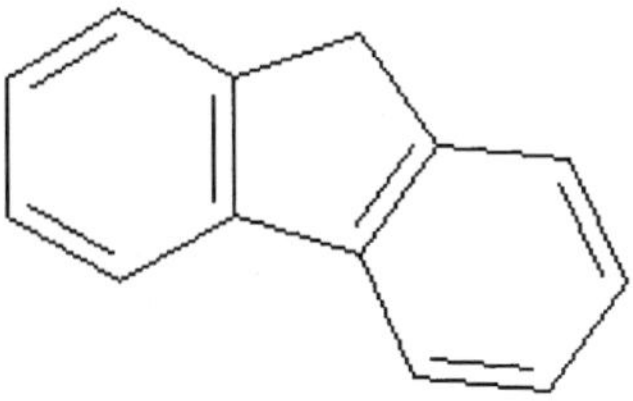

Fluorene

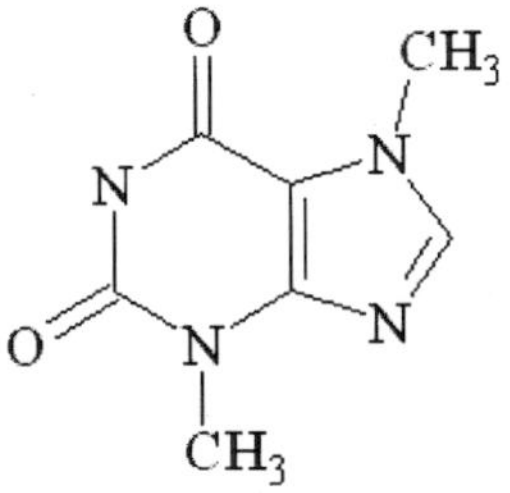

Theobromine

Fluorene is an unusual name in that the molecule doesn't contain the element fluorine! It gets its name from the fact it fluoresces under UV light. Similarly, *theobromine* doesn't contain any bromine. It's derived from cocoa trees (*Theobroma*), and is the bitter taste in dark chocolate. *Theos* actually means 'god' in Greek, and *broma* means 'food'. So, chocolate really is the food of the gods!

Gardenin

If you fancy a bit of gardenin', this is the molecule for you. In fact, there are many different *gardenins*, which are *flavones* extracted from *Gardenia lucida*, a plant from India [32]. The structure shown is for *gardenin A*, which forms yellow crystals.

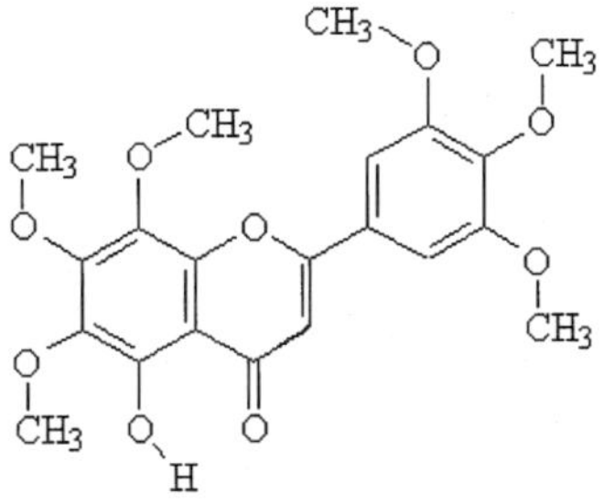

Uranate

UO_2^-

The various *uranium oxide* anions (UO_2^{2-}, UO_3^{2-}, UO_4^{2-}, *etc*) go by the glorious name of *uranates*. I wonder if unwanted reactions of these ions are called 'involuntary uranation'...?

And is *nickel uranate* what you'd need to 'spend a penny'? Related to this, *uranium nitrate* is also known as *uranyl nitrate*, which sounds like the entry fee for a gents' toilet after 8pm.

Conantokin

This chemical sounds like Conan the Barbarian has been smoking something he shouldn't... In fact it's a peptide neurotoxin found in the marine snail *Conus geographus*.

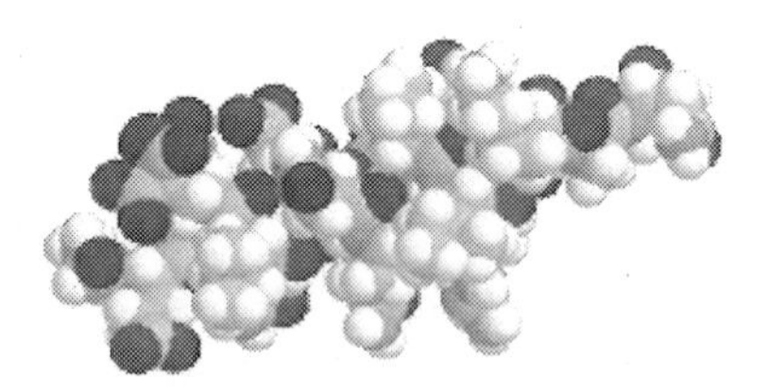

(Image created from 3D structure file [33]).

Researchers have found that some *conantokins* cause young mice to fall asleep, and older mice to become hyperactive, but they don't say what happens to middle-aged mice... It gets its name because it was isolated from *Conus* snails hence "con-". And, "*antok*" is a Filipino word which means "sleepy", which refers to its effect on young mice [34]. Apparently, there is also a related molecule called *contulakin*. "*Tulak*" is a Filipino term for "push". It seems that this molecule causes mice to be sluggish and thus, they had to be pushed.

Propellane and Cubane

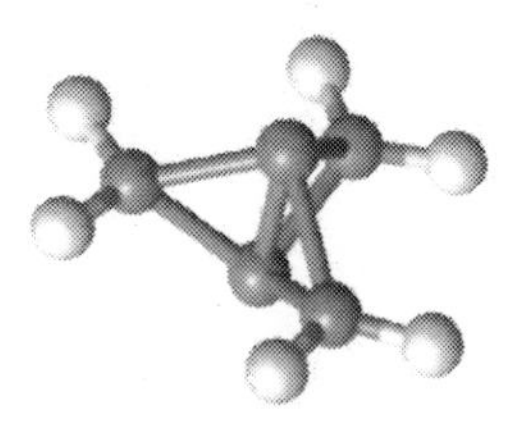

These two molecules are both named after their distinctive shapes. *Propellane* (left), C_5H_6, resembles a propeller [35], whereas *cubane* (right), C_8H_8, is a cube [36] (but doesn't come from Cuba). Other molecules that get their name from their geometric shapes are [5]: *dodecahedrane* $C_{20}H_{20}$, *prismane* C_6H_6, *spherands* and *hemispherands*, *squaric acid* $C_4H_2O_4$ and *deltic acid* $C_3H_2O_3$, *tetrahedranes* C_4H_4 and $C_{20}H_{36}$ and finally *twistane* $C_{10}H_{16}$.

Clitoriacetal

Apparently men find this molecule difficult to find... It gets its name from the root of the *Clitoria macrophylla* plant [37], and is a constituent of the Thai drug "Non-tai-yak" [38] which is used to treat respiratory disorders, including pulmonary tuberculosis and bronchitis, and also works as an insecticide.

Vaginatin

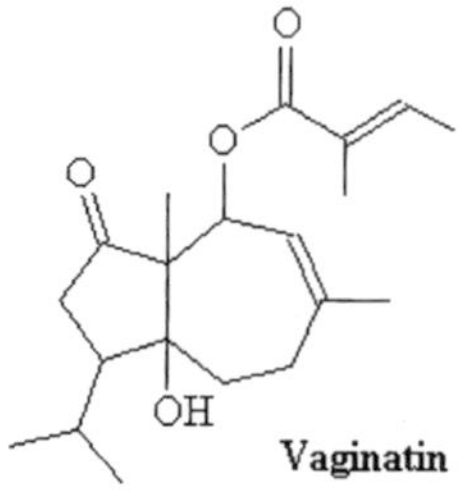

Vaginatin

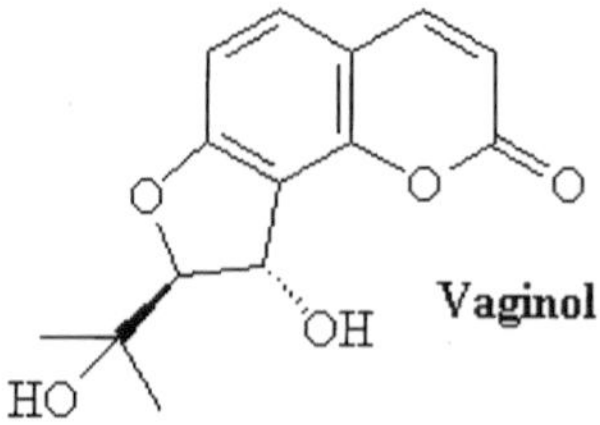

Vaginol

I know you can get most things nowadays in a tin, but this is getting silly! Actually it gets its name from the plant *Selinum Vaginatum*. A related molecule is *vaginol*, which also goes by the name *archangelicin*.

Anol

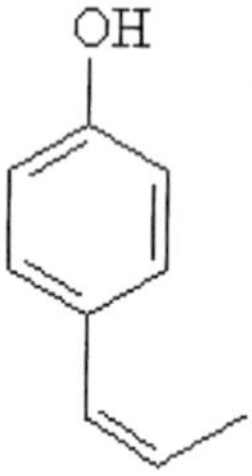

Anol is a synonym for 4-(1-*propenyl*)*phenol*, and it is apparently used in the flavour industry. Are compounds that bond strongly to this molecule called 'anolly retentive'?

Urospermal

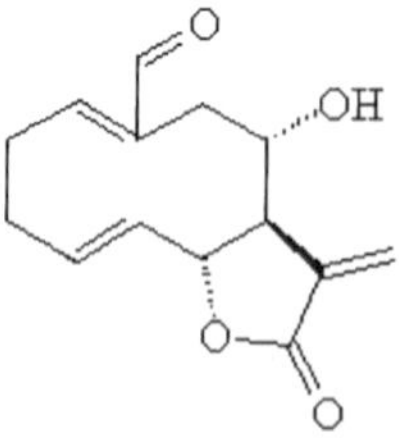

Is this an IVF clinic in Europe? In reality, it gets its name from being a constituent of the roots of the *Urospermum delachampii* plant.

Dogcollarane

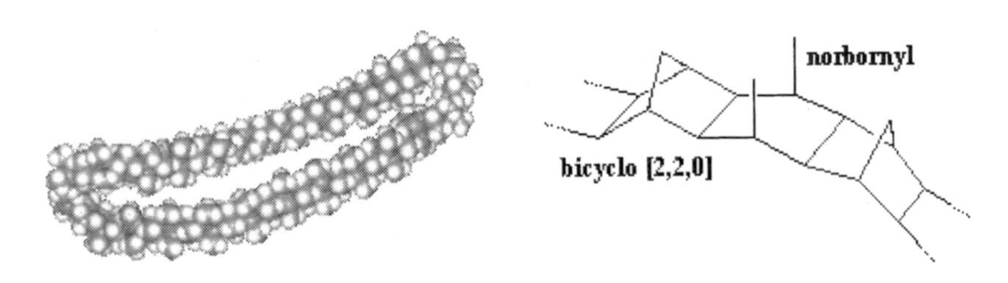

Dogcollaranes are a group of molecules made from alternating *bicyclo*[2,2,0] and *norbornyl* segments [39]. When there are 24 such components, the ends can be linked together to form a ring, which looks like a dog-collar. Unfortunately, although many of the intermediate structures have been made, none of the *dogcollaranes* have yet been synthesised. Shouldn't these molecules contain lead? (dog-lead, geddit?).

Antipain

Antipain is a protease inhibitor, which means it prevents proteins from being degraded. Despite its promising name, it is a very toxic compound, and it causes severe itch or pain (!) when contacted with the skin.

Its name is actually a contraction of *anti-papain*, since it inhibits the action of *papain*, an enzyme found in papayas that's used to treat bee and jellyfish stings.

Dinile

Why did the two cyanide groups go to see a psychiatrist?
...Because they were both 'in *dinile*'.
In fact, *dinile* is another name for *butanedinitrile* or *succinonitrile*, and is a waxy solid that if ingested forms cyanides in the body.

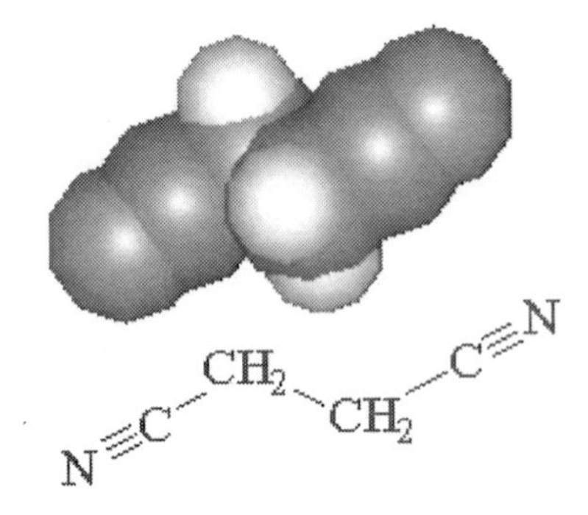

Butanal

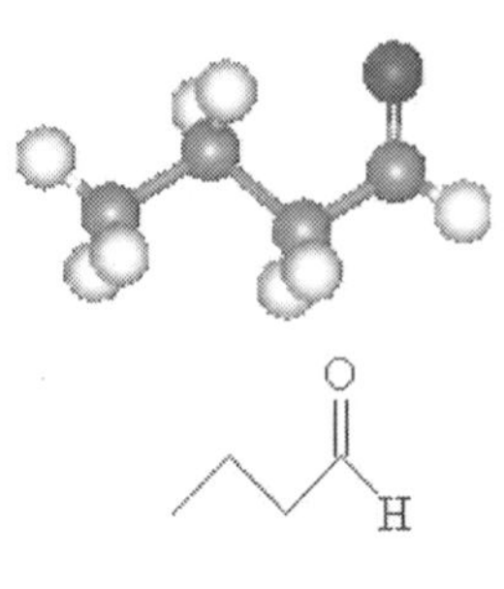

This molecule sounds better if it's hyphenated (but-anal), but it is actually quite a common aldehyde, also known as *butyraldehyde*.

Cacodyl

$(H_3C)_2As{-}As(CH_3)_2$

This molecule gets its name from the Greek *kakodes*, meaning 'stinking', as it has a really pungent smell of manure with a delicate hint of garlic. It is sometimes spelled *kakodyl*, but its correct name is *tetramethyl diarsenic*. Its main claim to fame is that it was one of the compounds worked on by Bunsen (of burner fame).

Angelic Acid

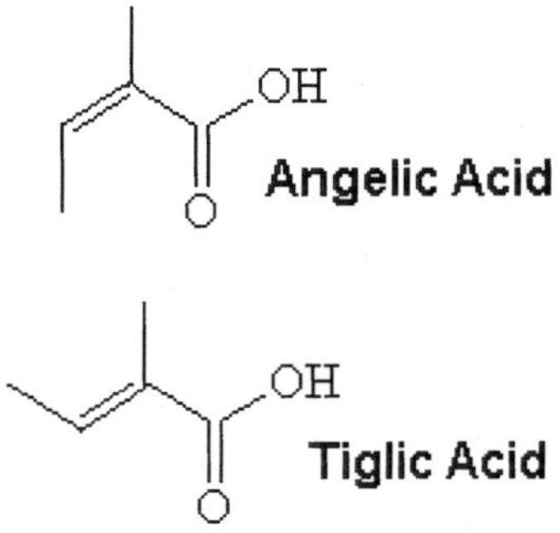

Angelic acid isn't very angelic at all - it's a defence substance for certain beetles. It gets its name from the Swedish plant Garden Angelica (*Archangelica officinalis*) from whose roots it was first obtained in the 1840s. Its proper name is *(Z)-2-methyl-2-butenoic acid*.

Angelic Acid

Tiglic Acid

The other isomer (*E*) is also a beetle defence substance and goes by the equally silly name of *tiglic acid* (from the plant *Croton tiglium*, which gets its name from *tilos*, Greek for diarrhea!).

Ciglitizone

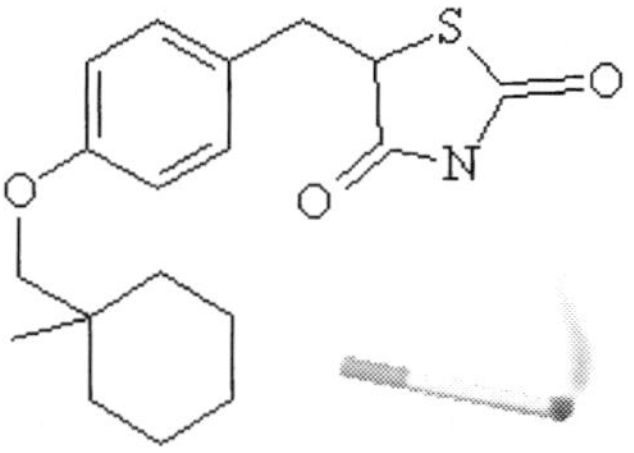

This molecule sounds like the places reserved for smokers to light up. Actually, *ciglitizone* is a member of a class of compounds that are used as anti-diabetics [40]. The drug *Avandia* (*rosiglitazone*), used to treat type II diabetes, is a member of this class of compounds.

Another related molecule is *troglitazone*, which I've mentioned for fans of cave-dwelling dwarfs.

Clitorin

I don't know much about *clitorin*, except that it's a *flavenol glycoside* [41] (make of that what you will), but I'd like to bet that it's touch sensitive...

Constipatic Acid

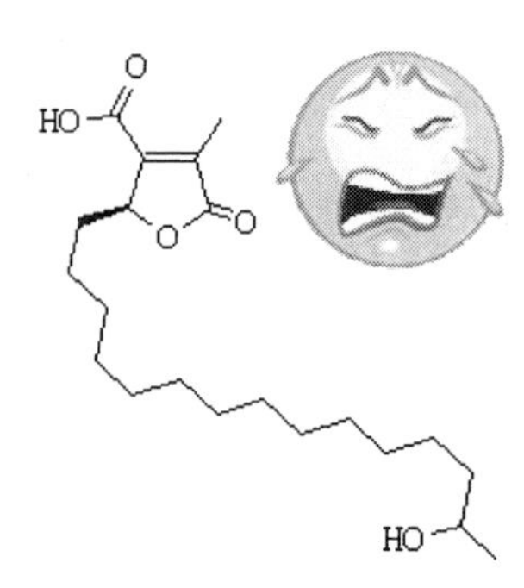

This is a constituent of some Australian lichens, including a fungus called *Parmelia constipata* which gave this molecule its wonderful name [42]. Derivatives of this molecule are *protoconstipatic acid, dehydroconstipatic acid* and *methyl constipatate.*

Fucol

L-Fucol

This sugar sounds like it doesn't do very much! Actually the *L-Fucol* form is obtained from the eggs of sea urchins, frog spawn and milk. The *L-fucol* form also goes by the name of *rhodeose* (yee-har!).

D-Fucol

Penguinone

This molecule gets its nickname from the similarity of its 2D structure to a penguin. The effect is slightly lost in the 3D model, though. It's real name is: 3,4,4,5-*tetramethylcyclohexa*-2,5-*dienone.*

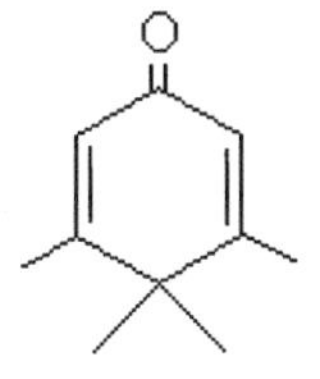

(Photo: Stan Shebs[43])

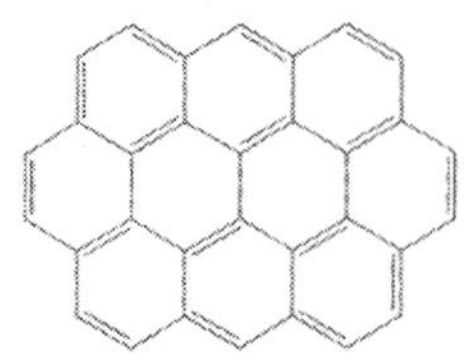

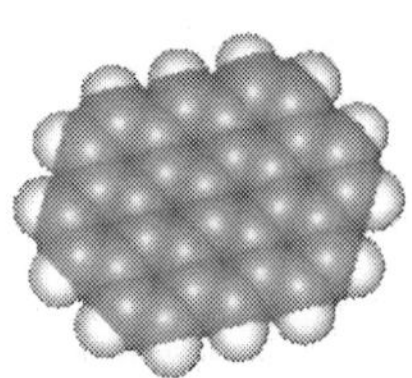

Ovalene

Ovalene is a polycyclic aromatic hydrocarbon [44], $C_{32}H_{14}$. Funnily enough it's oval-shaped... In the series of polycyclic aromatic hydrocarbons to which *ovalene* belongs, the next one is *circumanthracene*... (is it also known as *oy-vane?*).

Sepulchrate and Sarcophagene

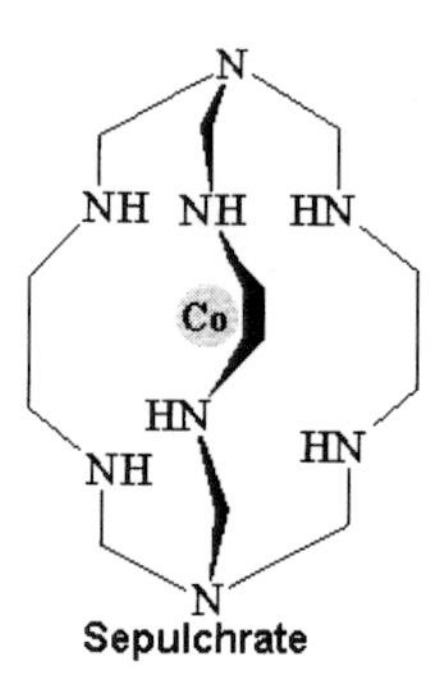

Sepulchrate

These spooky sounding molecules both have structures which wrap around and enclose metal atoms, such as cobalt, in a coffin-like cage. [45] Hence their funereal names.

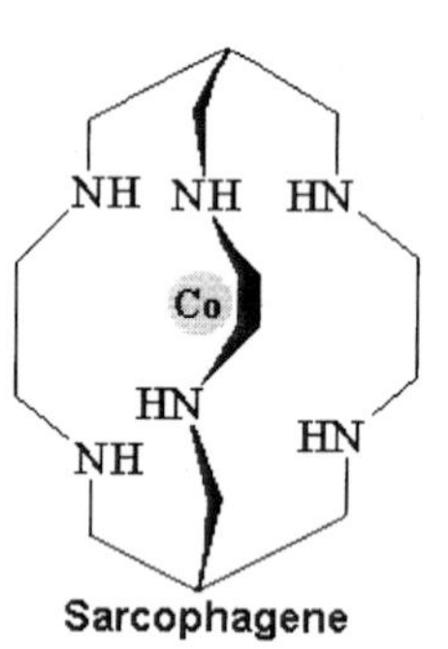

Sarcophagene

Pagodane

This $C_{20}H_{20}$ molecule gets its name because it resembles a Japanese pagoda[46] - well, two pagodas, back-to-back.

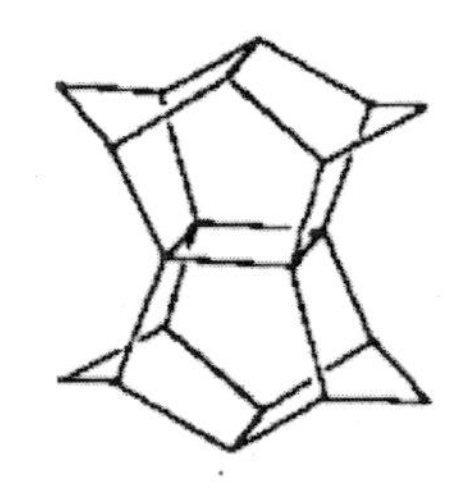

Left: The pagoda at the Kiyomizu-dera in Kyoto.
(Photo: David Monniaux[47]).

DEAD and DEADCAT

DEAD is actually the acronym for *diethyl azodicarboxylate*, which is an important reagent in the well-known Mitsunobu reaction which performs a stereospecific conversion of an alcohol to a primary amine.

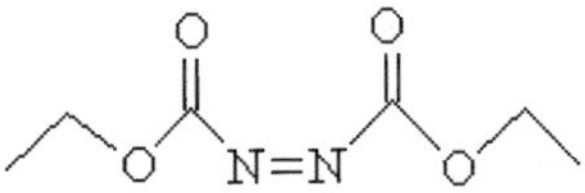

It's quite a good acronym, as *DEAD* is an orange liquid that's explosive, shock sensitive, light sensitive, toxic, a possible carcinogen or mutagen, and an eye, skin and respiratory irritant! A version of *diethyl azodicarboxylate* mixed with acid and *triphenylphosphine* has also been termed *DEADCAT*.

Lepidopterene and Biplanene

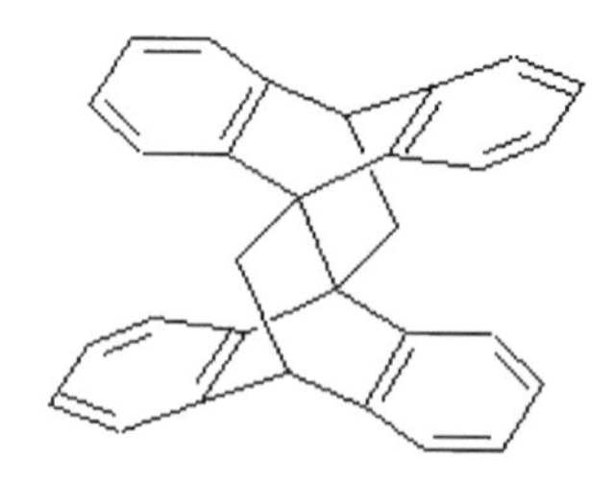

Lepidopterenes are a whole class of molecules named after their structural similarity to a butterfly. But when the two 'wings' are directly over one another, they look like a WW1 biplane, and so this group of molecules has been termed *biplanenes*[5].

Snoutene

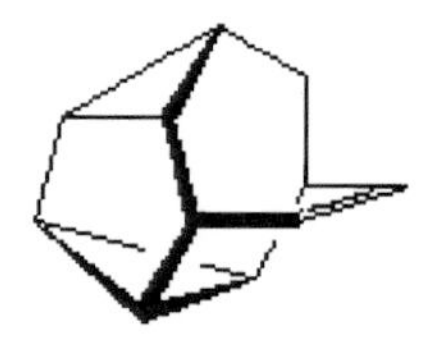

This strange looking molecule[48] resembles the nose or snout of an animal, but I don't know if it smells...

(Photo: David Monniaux[49]).

Crown Ethers, Lariat Ethers and BiBLEs

Both these ethers get their names from their distinctive shapes. *Crown ethers* look like crowns[50], whereas *lariat ethers* look like lassos [51], and are really just *crown ethers* with extended 'tails'.

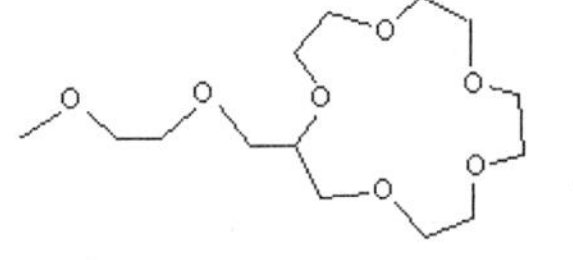

Some *lariat ethers* are so flexible that they can stick their tails into their rings (nice trick!), and so have been termed *ostrich complexes* [52], or *tail biters* [53]. *Lariat ethers* with two tails are called *bibrachial lariat ethers* (*bracchium* means 'arm'), and are abbreviated as *BiBLEs* [54].

Paddlane

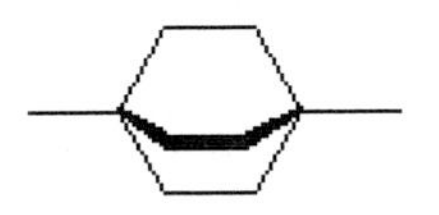

Paddlanes are molecules which have bicyclic *cyclohexane* units, which look a bit like the paddles on Mississippi steamboats [55].

Betweenanene (Screwene)

Betweenanenes are molecules which have a *trans* double bond shared between two *cycloalkanes*[56]. There is a whole family of them, depending upon the size of each ring. The one shown on the right is the [10,10] *betweenanene.* If there are two double-bonds linked together, the molecules are called *screwenes*[57], but this terminology isn't that popular, for obvious reasons!

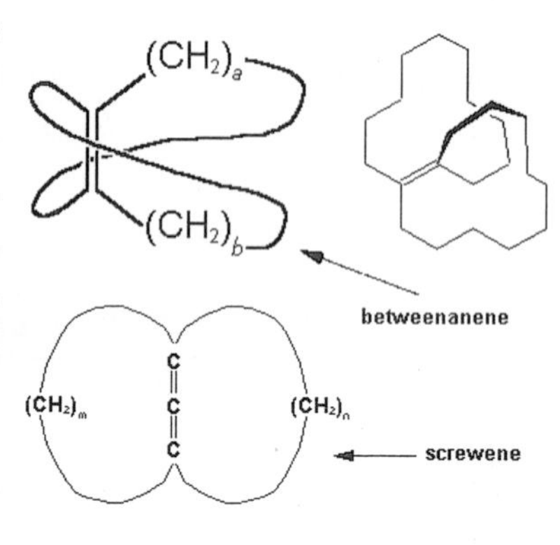

Furfuryl Furfurate

I bet Dr. Michael Hfuhruhurr could use some *furfuryl furfurate* about now...

(Copyright Getty Images, Warner brothers, taken from the film: The Man with 2 Brains, used with permission).

A ridiculously-named molecule, about which I know virtually nothing, although I'm told it's quite smelly and may be used as a vapour phase polymerisation inhibitor. It got its name from the Latin *furfur,* meaning "bran" (the source of the compound). A related molecule, *furfural alcohol* is apparently used in the fabrication process of the Reinforced Carbon-Carbon (RCC) sections used in the space shuttle.

George and Bi-George

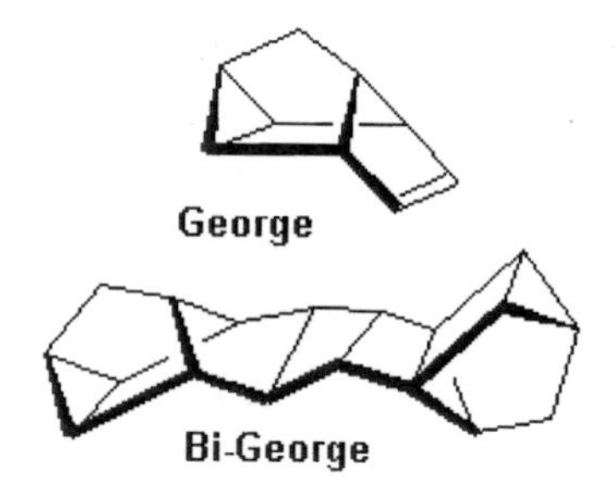

The story goes [5,58] that when undergraduate James Carnahan achieved the synthesis of a new cage structure at Columbia University, he asked his supervisor Prof Katz to suggest a name for it. Since trivial names are often arbitrary, he suggested *George*. When *George* was heated with a Rh catalyst, it dimerised to produce *Bi-George.*

Eurekamic Acid

Eureka! was supposedly the exclamation used by Archimedes when he found something interesting in his bath water. It means 'I have found it', and so when researchers at May & Baker discovered this acid, they felt it was such a 'Eureka moment' that they named the molecule after it [59].

Rudolphomycin and Rednose

Rudolphomycin

Rednose

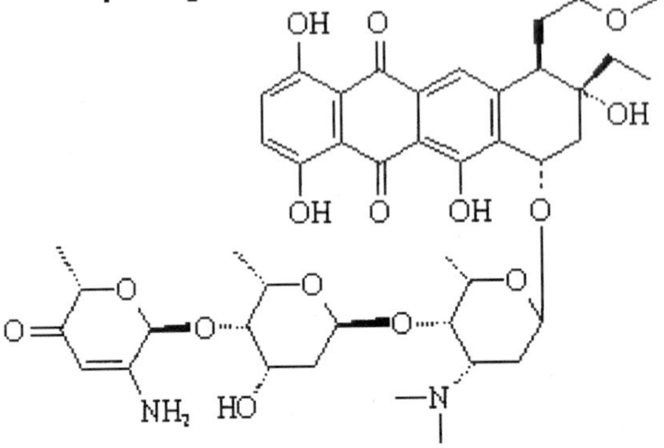

Rudolphomycin is an antitumor and antibiotic compound. It was named following a series of such molecules derived from *bohemic acid complex* - which was given its name because the discoverer, Donald Nettleton [60], was an avid opera fan, and called it after the Puccini opera *La Bohème*.

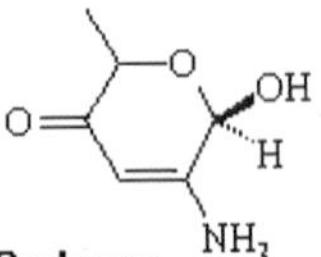

Derivatives of this molecule were then given names from characters in the opera, such as *mussettamycin* and *marcellomycin*, after Musetta and Marcello, *mimimycin* (after Mimi), *collinemycin* (Colline), *alcindoromycin* (Alcindoro), *schaunardimycin* (Schaunard), and finally *rudolphomycin* after the character Rudolph [61]. On degradation of *rudolphomycin*, a new sugar was obtained, which was christened *rednose* [62]. This rather silly name was probably allowed to stand since the paper was submitted to the journal very close to Christmas in 1978, and the journal editor probably had the Yuletide spirit! [5]

Catherine

Or *Cathy* to its friends? I don't know much about this molecule except that its name comes from the plant *Catharanthus roseus* [63]. I'm surprised, though, that it's not wheel shaped.

Complicatic Acid

This molecule didn't get its name because it was complicated to make, but rather from the plant *Stereum complicatum* from which it was isolated [64].

Ptelefolone

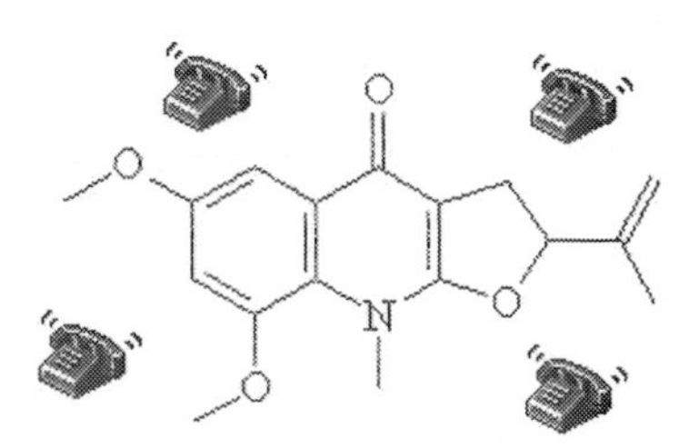

This molecule rings a bell! It's an alkaloid [65] that gets its name from the hop tree *Ptelea trifoliata* from which it was first isolated.

Pterodactyladiene

This is a group of molecules that resemble the ancient flying reptiles [66]. The R groups can be altered to give different sized 'heads' or 'tails'.

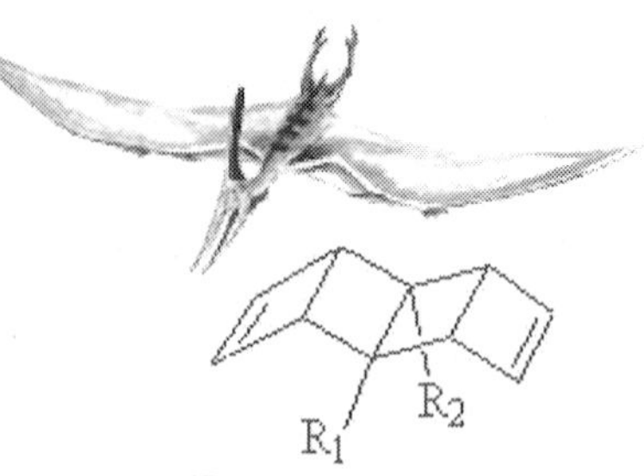

Pteranodon[67] and *pterodactyladiene.*

Miazole and Urazole

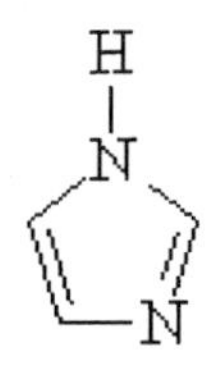

Miazole

If you pronounce the 'a' as in 'cat', and the 'z' as an 's', then you get the classic chemistry joke: What's the difference between *miazole* and *urazole*? The size of the ring...

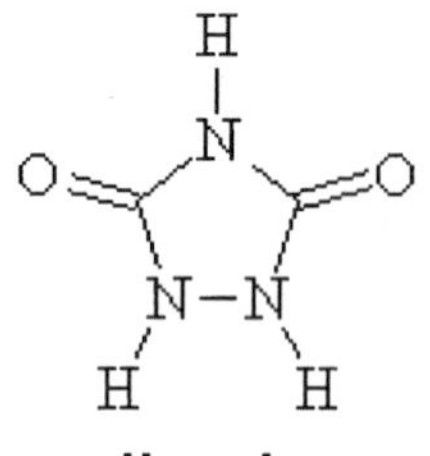

Urazole

And shouldn't there be a '*herazole*', a '*theirazole*' and an '*ourazole*' to get a complete bunch of *azoles*? Actually, the proper name for *miazole* is *imidazol*, but that spoils the joke a bit.

Ethyl Lactate

This is another standard undergraduate chemistry joke, based around the fact that *ethyl* sounds like a common female name. "How do you make *Ethyl lactate...?*" (I'll leave you to make up your own answer...but you could try *domperidone*, see above).
Other names involving *Ethyl,* such as *Ethyl palpitate, Ethyl fornicate* and the spinster *Ethyl celibate* also make good jokes, but unfortunately the corresponding acids (*palpitic, fornic* and *celibatic*) don't exist. Similarly, is *copper tartrate* what policewomen get paid to impersonate prostitutes?

Cristane

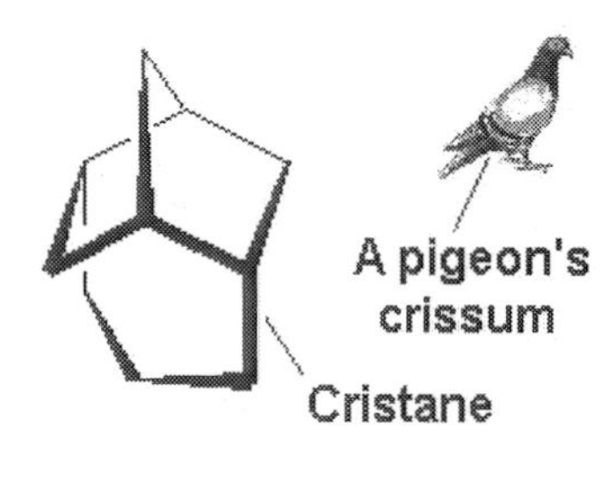

So what's so amusing about *cristane?* Well, for the non-biologists amongst you, a *crissum* is the name given to the area near the anus of a bird! *Tricyclo*[5.3.0.0]*decane* was given the nickname *cristane*[68] since on the evening it was first discovered, a window in the lab was left open.
A pigeon got into the lab overnight and did what pigeons do - all over the lab and equipment. The clean-up crew named the new molecule in honour of the part of the bird's anatomy that had created the mess[5].

Birdcage

This molecule is so called because it, um, resembles a birdcage [69]. Maybe it should have been used to capture the aberrant pigeon from *cristane*, above...

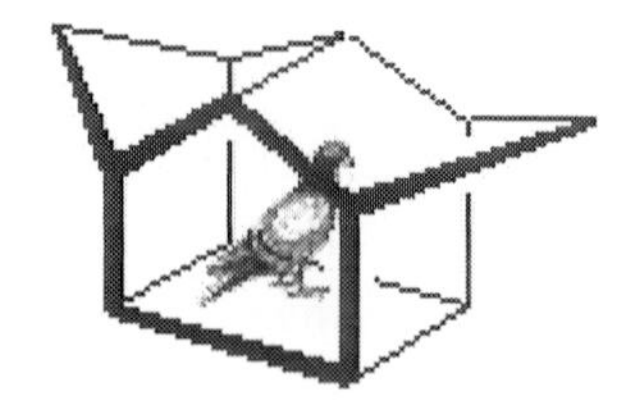

Cornerstone

Ok, this is a bit of a cheat, since its real name is β-*corrnorsterone*, but it's known as *cornerstone* by all those that work with it. It got its name since it's a *ketone* with a *norsteroid* structure (hence *norsterone*) and its discoverer, Robert Woodward, thought it might eventually be possible to transform it into a *corrinoid*.

Apolloane and Rocketene

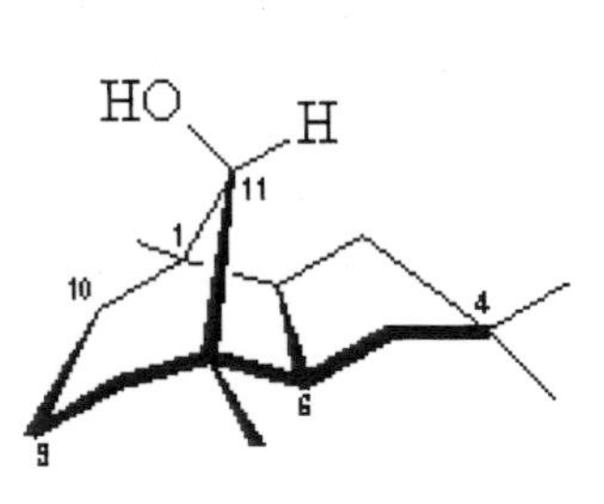

Apolloane-11-ol

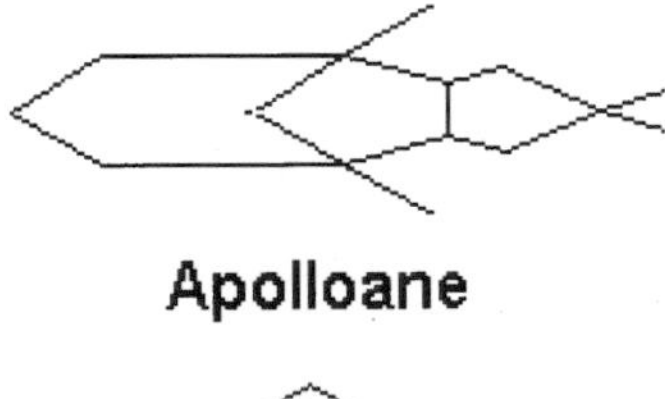

Apolloane

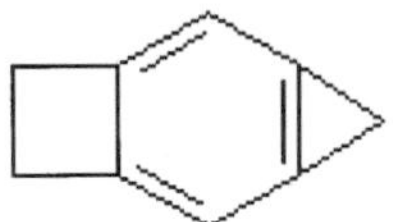

Rocketene

Apollo 11 launches
(Copyright Getty Images, NASA, reproduced with permission).

Apolloane was created at the same time as the Apollo 11 moon landings. When drawn as a flat diagram the structure bears a striking similarity to a rocket, with side fins and exhaust. And if an OH is added to carbon 11, we get *apolloane*-11-*ol*. Apparently, Neil Armstrong's personal memorabilia include a reprint of the chemistry publication which named it [5,70].

On a similar theme, *rocketene* was also named for its structural resemblance to a rocket.

Adamsite

This molecule was named after the renowned chemist Roger Adams of the University of Illinois [71]. It's actually a chemical warfare agent, and is 'a damn sight' better at killing people than most other molecules.

Manxane and Manxine

Manxane resembles the coat of arms of the Isle of Man (called a *triskelion*) which consists of 3 armoured legs in a circle. *Bicyclo*[3.3.3]*undecane* was named *manxane* since it closely resembles this Manx emblem [5,72].

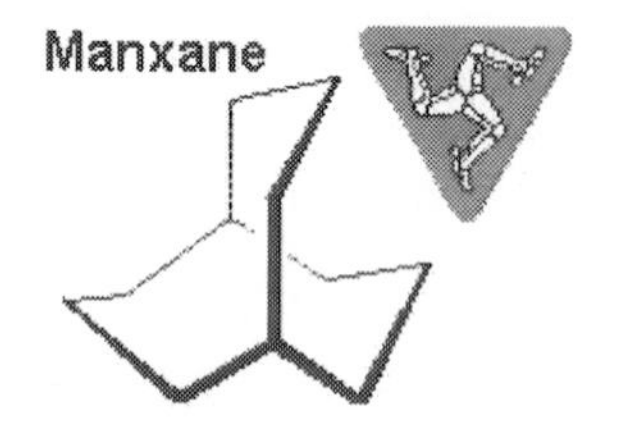

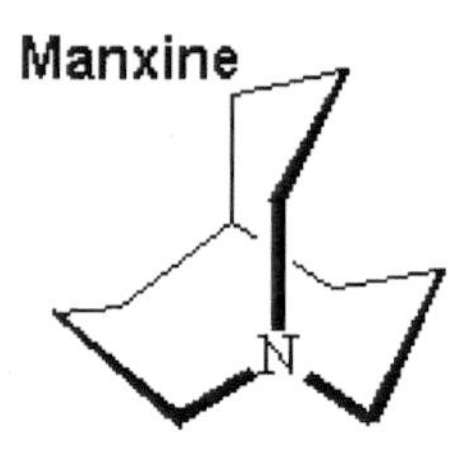

Later on, a group of researchers at the University of Illinois created an *amine* analog of *manxane* with a bridgehead nitrogen, and so called it *manxine* [73]. Professor Leonard, who created this molecule, thought *manxine* sounded like a girl's name - so we now have the male (*manxane*) and female (*manxine*) versions of the molecule, with the difference being what is situated between their legs!

BON-BONs

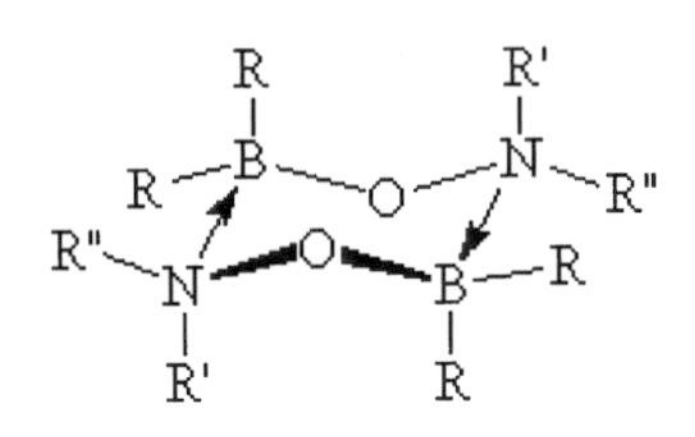

These ring structures are not what makes French sweets taste, well, sweet. Heterocyclic dimers like the one shown in the picture (where you vary the R, R' and R" groups) are named from the fact that the ring atoms in sequence spell out BON-BON [74].

Performic Acid

An actor's favourite chemical? As you might expect from a *per-acid*, it's a very strong oxidising agent, and always puts on a per-fect per-formance!

R-CMP

As anyone from Canada will know, R.C.M.P. are the initials for the Royal Canadian Mounted Police, but this molecule isn't their emblem. *R-CMP* is actually short for *R-cytidine monophosphate*, and is actually a component of RNA.

Gibberelic Acid

Gibberelic acid isn't a psychotropic drug that makes you gibber insanely like a monkey...it's actually one of a number of *gibberelins*, which are plant hormones which control various aspects of plant growth.

Jesterone

This playful and mischievous molecule is found in a fungus, *Pestalotiopsis jesteri*, which lives inside yew trees [75].

Bellendine

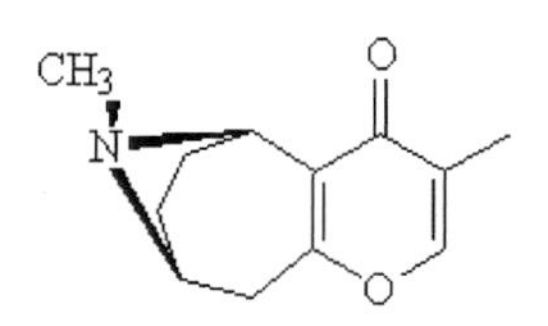

For non-UK readers, I won't tell you what 'bell-end' is slang for in the UK, but a clue might be that this molecule is extracted from the flower *Bellenda Montana* [76], which should have a purple head.

Sodamide

This is the shorthand name given to the common chemical, *sodium amide*, $NaNH_2$. It sounds like it belongs in close proximity to other molecules in this list, such as *arsole*, *anol*, *skatole*, and maybe *fruticolone*... But I'm not letting it get anywhere near *miazole*!

Darlingine

This molecule is lovingly extracted from the Brown Silky Oak tree, *Darlingia darlingiana*[77]. It hasn't been analyzed biologically yet, but may have activities similar to other *tropane alkaloids*, such as muscle contraction and stimulation... (But it only works if you treat it nicely).

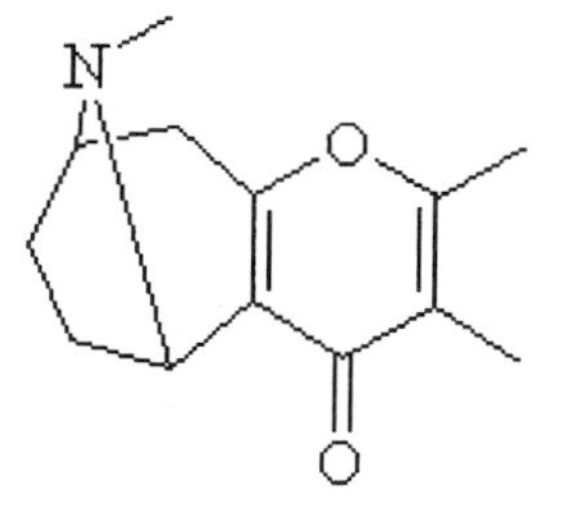

Germylene

(Photo: Ng Paik Mui).

The GeH_2 radical is called *germylene*, which is similar to a UK antiseptic ointment called *Germolene*. I doubt that *Germolene* contains *germylene*, though, as GeH_2 is very toxic.

Piano Stool

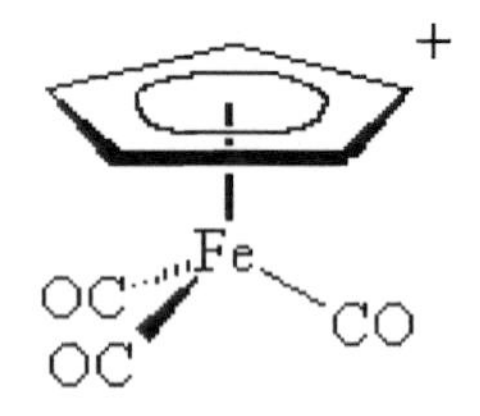

These are a group of molecules made from a transition metal bonded to a *cyclopentadienyl* ligand, so that they resemble a 3-legged piano stool [78]. I don't know if molecules have been made with other numbers of 'legs', such as milking stools, *etc.*

Trunkamide

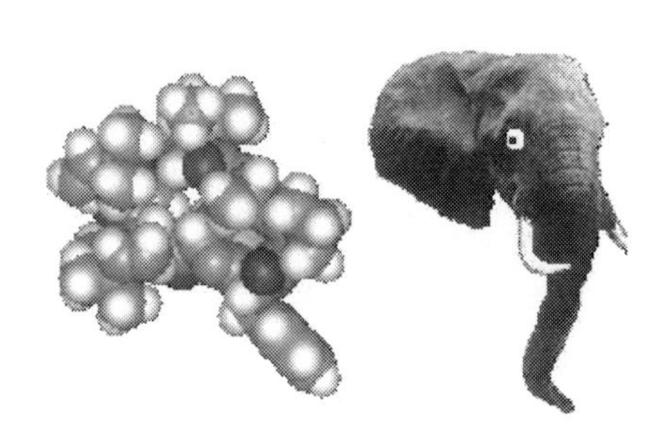

Trunkamide has nothing to do with elephants, although its spacefill structure does look a bit like an elephant. It was isolated from a sea squirt living in Little Trunk Reef (part of the Great Barrier Reef off Australia) [79], and is reported to have anti-tumor properties.

Diurea

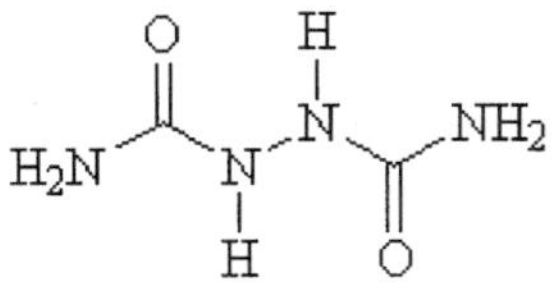

As you might expect, this molecule and its derivatives are often used as a fertiliser, being splattered liberally around fields of crops. It's also known as *biurea*, but its proper chemical name is *N,N'-dicarbamoylhydrazine.*

It's also sometimes used as flow improver in paints and greases. So next time you paint your house you can tell people you're covering it with *diurea.*

dUMP and RUMP

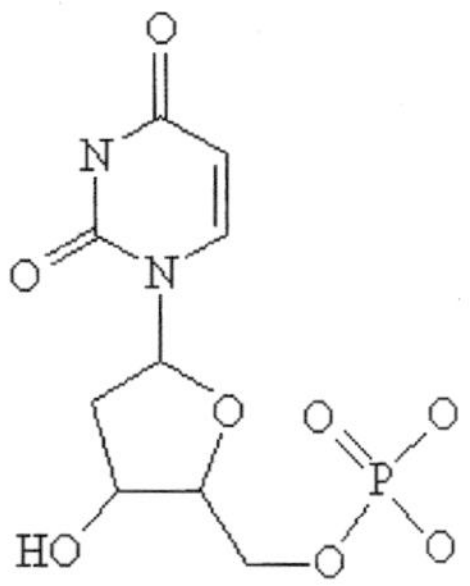

Maybe *dUMP* is the molecule into which all the waste atoms are thrown? In fact it's the acronym for 2'*-deoxyuridine-*5'*-monophosphate,* and is an RNA transcription subunit - or a part of the molecule that makes proteins, and is one of the building blocks of DNA. Maybe *dUMP* makes so-called junk-DNA? *dUMP* is a deoxygenated form of *RUMP,* which is another of the bases found in (near the bottom of?) RNA.

BARF

There are two molecules that are called by their shorthand name of *BARF*. The first, often written as *BAr^F*, is a halide abstracting reagent $B[3,5\text{-}(CF_3)_2C_6H_3]_4^-$, *i.e.* two CF_3 groups on each *phenyl*, and four of those *phenyls* on a boron.

The second is written *BARF*, and is the shorthand for *tripentafluorophenylborane* (B-Ar-F), see structure on the right. It is mainly used as a strong Lewis acid to abstract a *methyl* group in the reaction to make a highly active *ethylene* polymerization catalyst. So you really can barf into a plastic bag!

Small-breasted dog

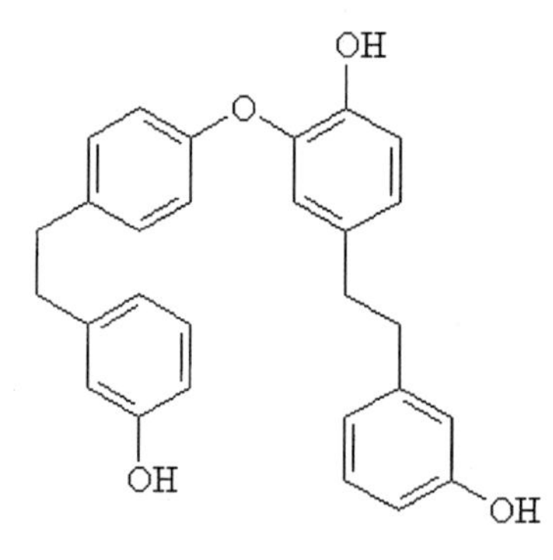

Yes, it really is called that, but in Spanish! The molecule is actually named *perrottetin-a* which (almost) literally means "small-breasted-dog" (*perro* = dog, *tetita* = small breast).

The molecule gets its name from the liverwort plant from which it is extracted [80], *Hepatica Radula perrottetii*, but I don't know why this plant is named after a small breasted dog.

Centaureidin

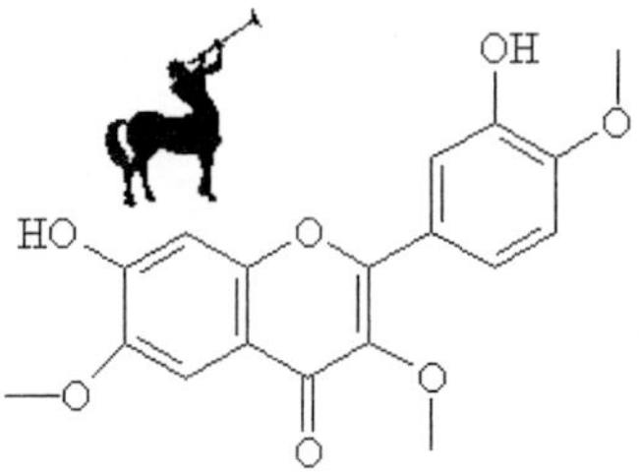

Centaurs are mythical creatures which are half horse, half human. The molecule *centaureidin* got its name, not because its inventors were horsing around, but because it was extracted from a cornflower called *Centaurea corcubionensis*. It was discovered along with a molecule named *centaurein*, found in the same plant [81].

OH

HO

O

O

Sandwicensin

HO

O

O

O

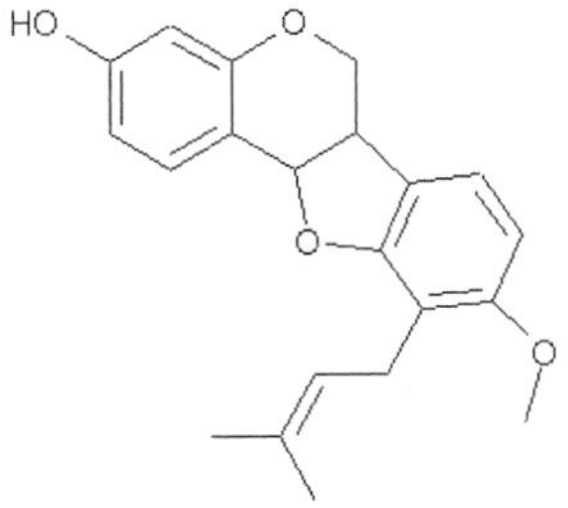

John Montagu, The Fourth Earl of Sandwich, was a notorious gambler who would often go from pub to pub in London on gambling marathons. To satisfy his hunger while continuing to gamble, he would order slices of meat between two pieces of bread - thus, was the sandwich born. But how about *sandwicensin*? It's a cytotoxin, isolated from a soggy old sponge, so I guess that it must be somewhat less than appetizing [82].

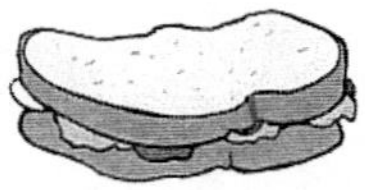

Magic Acid

'Magic Acid' is the nickname given to one of the strongest of the inorganic 'superacids'.

It is made by mixing together *antimony pentafluoride* (SbF_5) and *fluorosulfonic acid* (HSO_3F), and it is so strong (pK_a = -20) that it is capable of protonating even saturated alkanes, like *methane*, to produce *carbonium ions*. On a similar theme, *magic methyl* is the name given to *methyl fluorosulfonate* (F-SO_2-OCH_3) due to its extreme methylating power [83].

Megatomic Acid

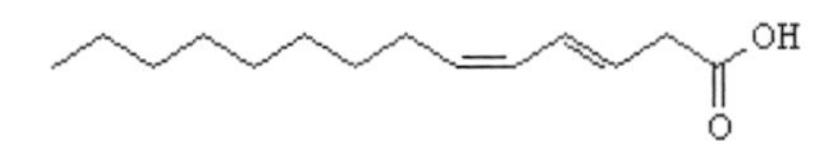

This molecule has nothing to do with nuclear explosions, and neither is it the magic formula that creates a superhero. But it is in fact named after the black carpet beetle *Attagenus megatoma* (*Fabricius*), in which it is the principle component of the beetle's sex attractant [84]. Its proper name is (3*E*,5*Z*)-3,5-*tetradecadienoic acid*.

Left: Looks like someone's just synthesised *megatomic acid*...!
(Copyright Getty Images, used with permission, Photo: Chad Baker).

Grasshopper ketone

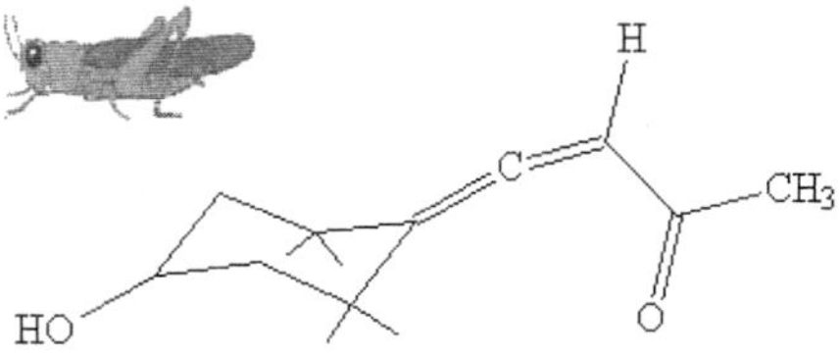

I'm guessing that this molecule got its name as a result of laziness. It's extracted from the defensive secretions of the flightless grasshopper *Romalea microptera*.

I assume that after spending hours in the field, annoying the grasshoppers, and then catching them and 'milking them', the scientists involved were too tired to think of a proper IUPAC name, so they came up with an inspired name - *grasshopper ketone* [85].

BCNU

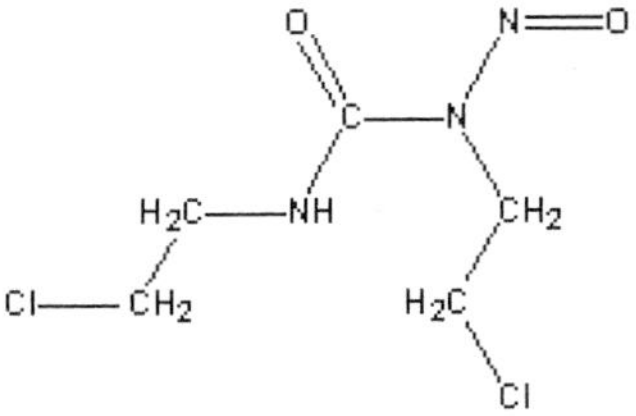

1,3-*bis*(2-*chloroethyl*)-1-*nitroso-urea* (also known as *carmustine*) has got quite an appropriate acronym, *BCNU* (be seein' you...), since in early medical studies it was found to be so toxic, it killed the patient! This isn't surprising, as it's related to *mustard gas*.

It is highly carcinogenic, causing tumors in rats, mice, rabbits, and probably humans as well. Ironically, it is actually used as a treatment for brain cancer and other diseases, such as Hodgkin's lymphoma [86].

SEX

SEX is the official abbreviation of *sodium ethyl xanthate,* which is a flotation agent used in the mining industry. Apparently you can get *SEX* in both solid and liquid forms (should that be hard *SEX* and wet *SEX*?).
According to Australia's National Industrial Chemicals Notification and Assessment Scheme [87] signs of high exposure to *SEX* include "dizziness, tremors, difficulty breathing, blurred vision, headaches, vomiting and death". Sound familiar...?
On a related note, there's another flotation reagent, *KAX, potassium amyl xanthate,* which has the same function, and the same smell.

Austin

Mike Myers as Austin Powers, in the film 'The Spy who Shagged me'. (Copyright Getty Images, used with permission).

Austin has nothing to do with Austin cars, Austin Texas, or even Austin Powers...it's actually a mycotoxin, which comes from the fungus *Aspergillus ustus* which grows on black-eyed peas [88]. The name is a contraction of *A. ust* plus '*-in*'.

Pantolactone

This molecule sounds like it belongs in underwear, or on stage in a pantomime. As you might expect, it is used as a reagent in the synthesis of *CAMP* ligands (*cis-2-(aminomethyl)-1-carboxycyclopropane*), and is found in cocoa [89].

Technetium Cow

'Cow' terminology comes from the nuclear industry, and it has nothing to do with the cattle that live near nuclear power plants. A radionuclide, such as ^{99}Mo (as its ammonium salt), is stored in a column, called a 'cow'.

A cow, but not a technetium one [90]

Its decay product, technetium-99m, is continually produced, and it can be flushed out of the cow column in a process called 'milking the cow'. The *technetium cow* isotope is then used in bone scans, and has a 6-hour half-life. On a related theme, *molybdic anhydride* (MoO_3) is often referred to as 'Moo'.

Erectone

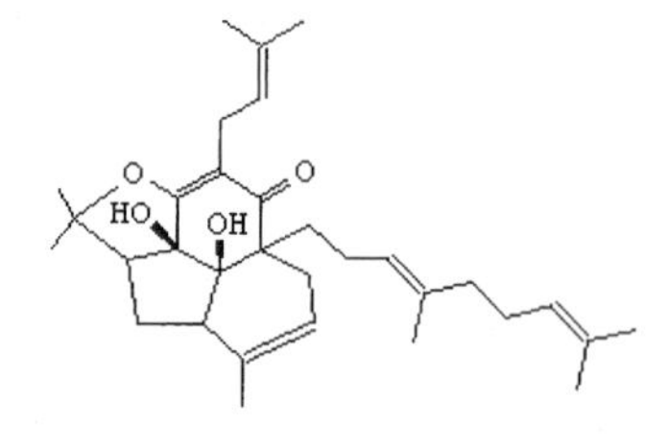

No, this isn't one of the ingredients in *Viagra*...but is actually one of a group of compounds extracted from the Japanese/Chinese herb *Hypericum erectum,* which is often used in traditional Chinese medicine to treat arthritis, rheumatism, and as an astringent [91].

Abiguene

Ambiguenes are cytotoxic fungicidal *indole alkaloids* that are extracted from blue-green algae (*Ficherellu Ambigua*) [92]. As many as 7 are known. Although it could just be 5. Or 10. Or maybe they are a different type of molecule altogether...?

Lovenone

This lonely sounding molecule (*love-none*) is a cytotoxic agent isolated from the skin of a *dorid nudibranch* (of course) called *Adalaria loveni* which lives in the North Sea.

Dorid nudibranchs are shell-less marine molluscs that you'd expect to be extremely vulnerable to predation...but aren't since they have evolved nasty chemicals such as *lovenone* as a defence [93]. In this case, love really can kill...

Boldine

Too much *boldine* on the Starship Enterprise's menu? (Patrick Stewart as Jean-Luc Picard in Star Trek:TNG, Copyright Getty Images, used with permission, photographer: George Rose).

This is an alkaloid extracted from the Chilean *Boldo* shrub (*Peumus boldus*) [94]. It is a good antioxidant and can protect the liver, although no-one's mentioned hair-loss as a side-effect?

Inflatene

This molecule is isolated from soft coral (*Clavularia inflata*) and is apparently toxic to fish [95] - maybe it blows them up! Geddit?

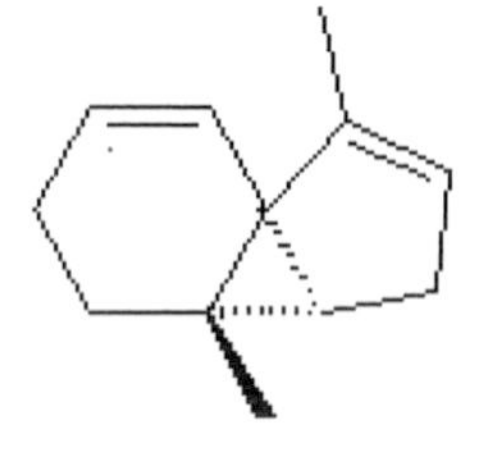

Bowtiediene

The molecule for formal occasions? This is another molecule named after its shape - although the preferred name is *spiropentadiene* [96].

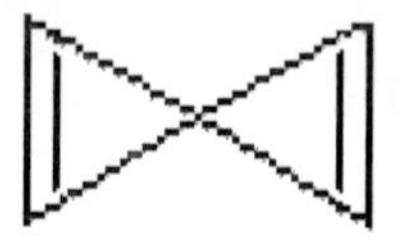

(R)-DICHED

This is the abbreviation for *(R,R)*-1,2-*dicyclohexyl*-1,2-*ethanediol* [97]. I wonder who the real *DIC-HED* is?

Prodigiozan

This molecule has a name that sounds like the Biblical 'prodigal son', who finally returned home. The structure shown is actually that of the related molecule, *prodigiosin*, since I can't find the structure of *prodigiozan*... well, not until it finally comes back here. Both molecules are antibiotic pigments produced by *Chromobacterium prodigiosum*, with antimicrobial and cytotoxic properties.

Arachidonic Acid

An avid user of *arachadonic acid*?
(Copyright Getty Images, used with permission, Photo: Mark Von Holden/WireImage).

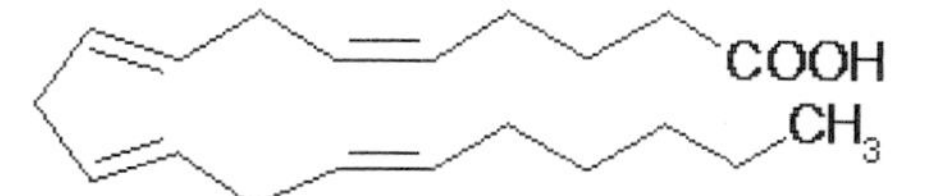

This molecule sounds like it has something to do with spiders, but it's actually made in the human body [98]. It is synthesised from *linoleic acid* and plays an important stage in the inflammatory process of the human body - some non-steroidal anti-inflammatory drugs are believed to work by inhibiting this stage.

Nobody has managed yet to artificially produce medical grade *arachadonic acid* (it's mainly used in infants) so the only source is rats' urine - it needs a day's worth of urine from 10,000 rats to produce a single dose! Now, that really is taking the p***!

Warfarin

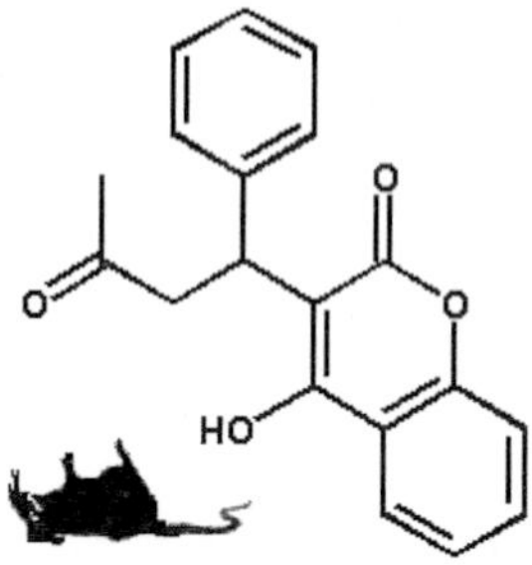

This molecule sounds like it could be a warfare agent, and it is...if you're a rat [99]. It's a rat poison which stops the blood clotting, so the rats bleed to death from the slightest injury. It also has medical uses in blood thinning and clot prevention. It gets its name since *WARFarin* was the first patentable product of the Wisconsin Alumni Research Foundation (WARF).

An interesting story about *warfarin* is that it may have been used to kill Stalin [100]. One medical side-effect of *warfarin* treatment is the dreaded 'purple toe syndrome', where small deposits of *cholesterol* break loose and flow into the blood vessels in the skin of the feet, which causes a blue-purple colour mainly in the big toe and may be painful.

Lunatoic Acid

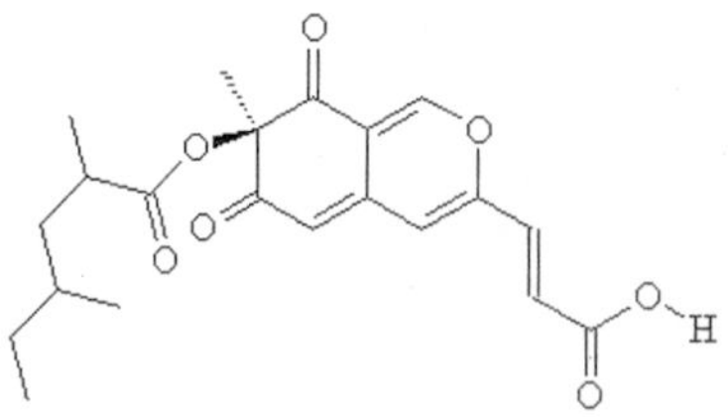

Lunatoic acid gets its name since it's isolated from the fungus *Cochliobolus lunatus* [101]. It is a good antibiotic and also causes fungi to shed their spores in a mad frenzy. Perhaps it kills bacteria by causing them to die by insanity, in the same way canine distemper kills animals.

DAMN

DAMN is the acronym for *diaminomaleonitrile,* which is a particularly nasty molecule containing two *cyanide* groups [102].

H_2N NH_2

NC CN

"Frankly my dear, I couldn't give a *diaminomaleonitrile...*"
(Clark Gable in 'Gone with the Wind', Copyright Getty Images, used with permission, photo: Clarence Sinclair Bull).

Allene

H H
C=C=C
H H

Allene is quite a sad molecule in Holland, since in Dutch it is called *alleen,* which simply means 'alone'. And if you add a *benzene* ring you'll get *benzo-allene* which means "I'm so lonely" in Dutch. Ahhhh....

Proton Sponge

According to the Aldrich Chemical Catalog, 1,8-*bis(dimethylamino)naphthalene* is a very strong base with weak nucleophilic character due to steric effects. Therefore it goes by the nickname *proton sponge,* since it mops up all available protons.

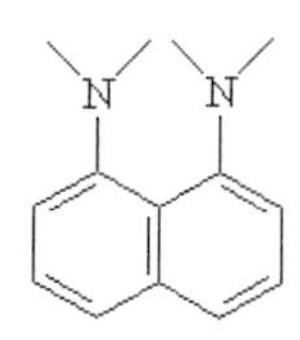

Mirasorvone

Left: Mira Sorvino, the actress.
(Copyright Getty Images, used with permission, Photo: Pascal Le Segretain).

This molecule forms part of the defensive chemistry of the recently named 'sunburst diving beetle' (*Thermonectus marmoratus*). The discoverers at Cornell University [103] named it in honour of the actress Mira Sorvino, who, as Dr Susan Tyler in the motion picture *Mimic*, successfully confronted the ultimate insect challenge.

Assoanine

This molecule gets its superb asinine name from the plant from which it is extracted, the gloriously named *Narcissus assoanus*!

Cumene

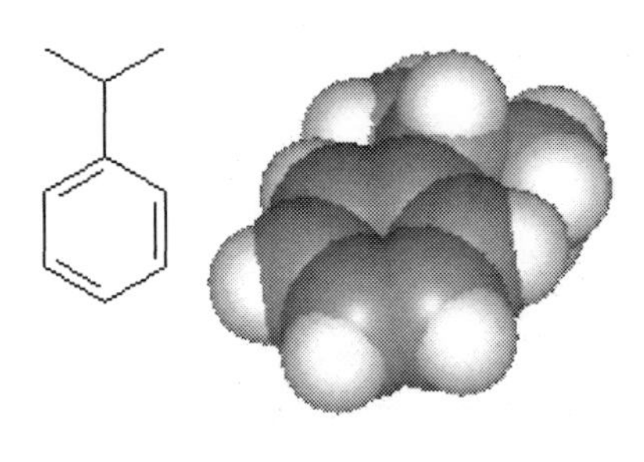

Luckily, this molecule is actually pronounced 'coo-mene', so as to avoid sticky problems when ordering it. It's a fairly standard organic solvent, with a distinctive odour, that is used to make resins, polycarbonate, synthetic fibres such as nylon, and other plastics.
Although pronounced the same, it has nothing to do with the Indian curry spice, cumin.

Wheelbarrow Molecule

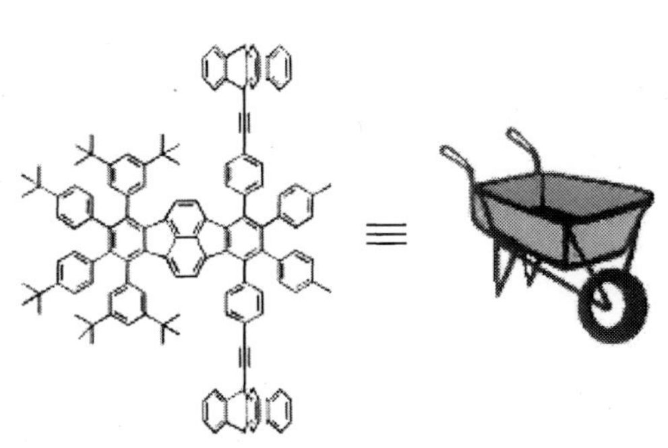

(Reproduced with permission from Elsevier, from ref.[104]).

Here's a molecule that has been designed to look like a wheelbarrow - well, why not? It doesn't seem to have a full name yet, so for now it's just called *wheelbarrow molecule.* What's next, a molecular lawn mower? Pruning shears?

Flea

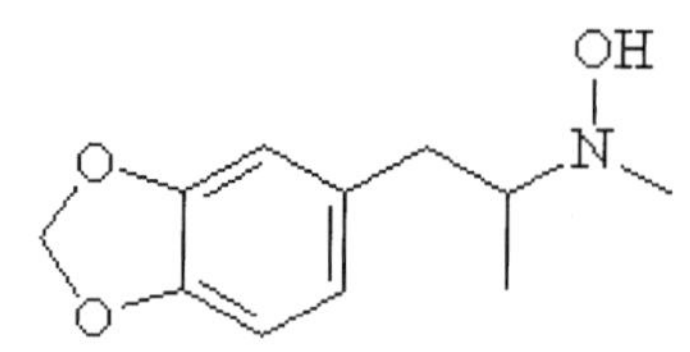

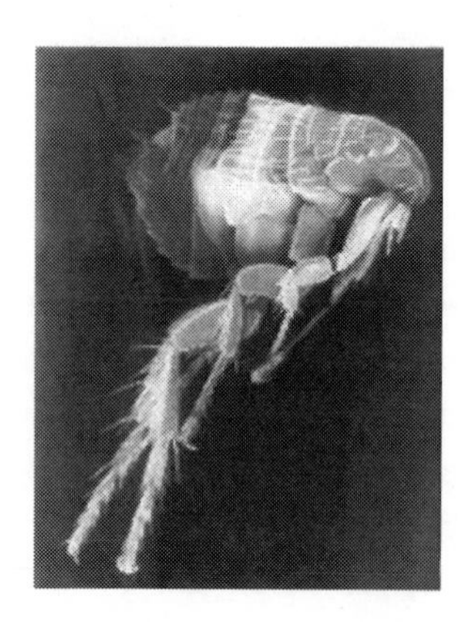

Right a SEM picture of a real flea [105].

This is the commonly used name for the amphetamine, *N-hydroxy-N-methyl-3,4-methylenedioxymethamphetamine* (the *N-hydroxylated* version of *MDMA* or *Ecstasy*). The origin of its name is a bit strange [106], and is related to the fact that a commonly-used code name for the parent compound, *MDMA* was *ADAM*. The 6-*Methyl* homologue was then called *MADAM*, and, following this pattern, the 6-*Fluoro* analogue was to be *FLADAM*. So, with the *N-hydroxy* analogue, the obvious choice was *HADAM*. But this brought to mind the classic description of Adam's earliest complaint, an infestation of fleas. The poem was short and direct: "Adam had 'em." So, in place of *HAD 'EM*, the term *FLEA* jumped into being.

Dopamine

Dopamine (pronounced dope-a-mean) is a neurotransmitter in the brain. It is connected to pleasurable sensations (feeling doped) and has been shown to be connected to drug abuse and addiction.

Maybe that's why Dopey from Snow White and the Seven Dwarves always had that silly grin on his face...

Syringic Acid

This molecule is named after the lilac plant, since the Latin name of lilac genus is *Syringa.* Lilac bushes possess hollow sticks which were used in ancient times to make flutes.

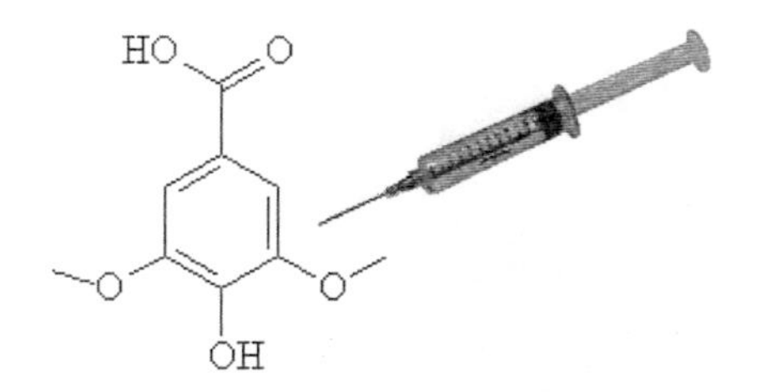

In fact, there is a kind of flute that is called *siringa* in Spanish. In Latin the meaning of the word *siringa* was extended to include hollow tubes made of any material, including metal. Later, when hollow needles began to be used to inject liquids in the body, quite naturally they were called syringes. Funnily enough, *syringic acid* can be found in blueberry plants, which, in Latin, are called *Vaccinium.* Quite a coincidence! On a related theme, there is also a *vaccenic acid* (*Z-11-octadecenoic acid,* also known as *asclepic acid*) although its structure is not related to that of *syringic acid.* Another interesting fact is that *syringic acid* is a component of red wine, and traces of this molecule found in Egyptian jars from King Tut's tomb show that the ancient Egyptians used to drink red wine [107]. Funnily enough, since 5 other wine jars did not contain *syringic acid,* this shows that they enjoyed white wine as well.

Tortuosine

Tortuosine - Viagra for tortoises?
(Copyright Getty Images, used with permission. Photographer: Mark Moffett)

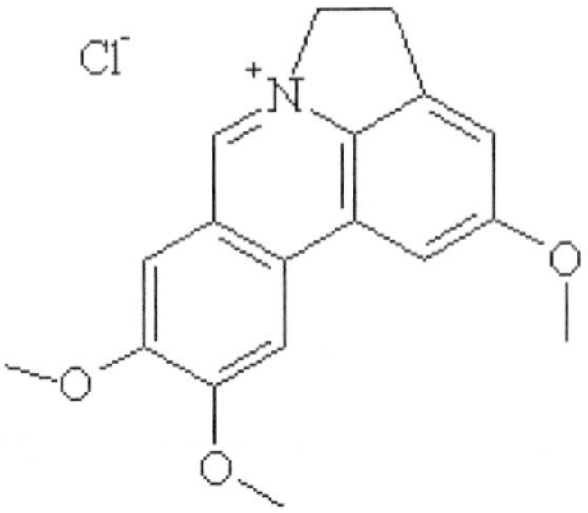

This molecule is an alkaloid extracted from the plant *Narcissus tortuosus*, but I bet it was extracted very sloooowwwllly. In fact, this naturally occurring organic compound, as well as *assoanine* (see above) had plant-derived names that were so compelling that Lee Flippin designed and executed total syntheses of them just for the fun of it! [108]

BIG CHAP

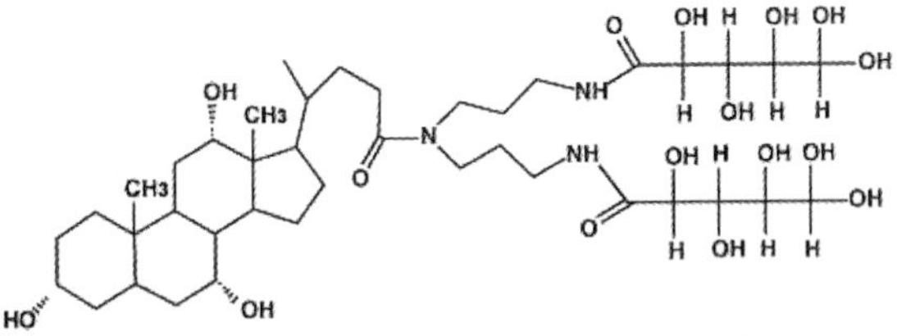

This wonderfully named molecule has nothing to do with reproduction, but is actually a detergent which has the official name of *N,N-bis(3-D-gluconamidopropyl)cholamide* [109]. Apparently the molecule has reduced electrostatic interactions that prevent your *BIG CHAP* getting stuck in a chromatography column.

NanoPutian Molecules

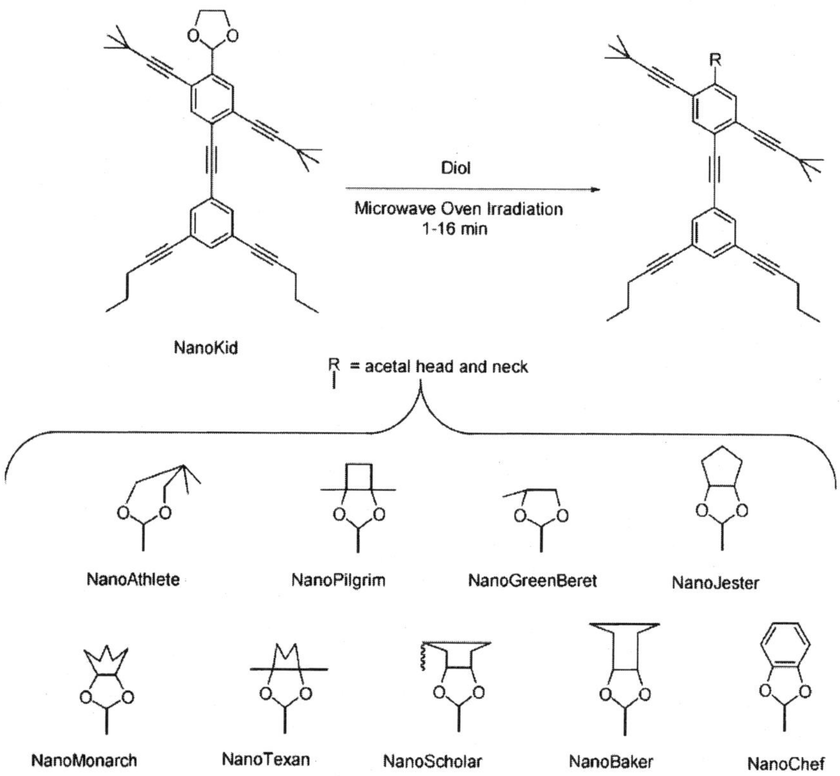

Diol

Microwave Oven Irradiation
1-16 min

NanoKid

R = acetal head and neck

NanoAthlete NanoPilgrim NanoGreenBeret NanoJester

NanoMonarch NanoTexan NanoScholar NanoBaker NanoChef

And today's award for the 'How did they possibly get a grant to do *that?*' paper, goes to two chemists from Rice University in Texas. Their paper [110] concerns making anthropomorphic molecules - *i.e.* molecules that look like humans...but why anyone would want to do this I don't know... The molecules have been named *NanoPutians*, after the little men from Lilliput in the book "*Gulliver's Travels*". They come in many forms - the basic building block is the *NanoKid*, and from this other variants can be made, such as *NanoAthlete* and *NanoBaker* and even *NanoBalletDancer* (see below).

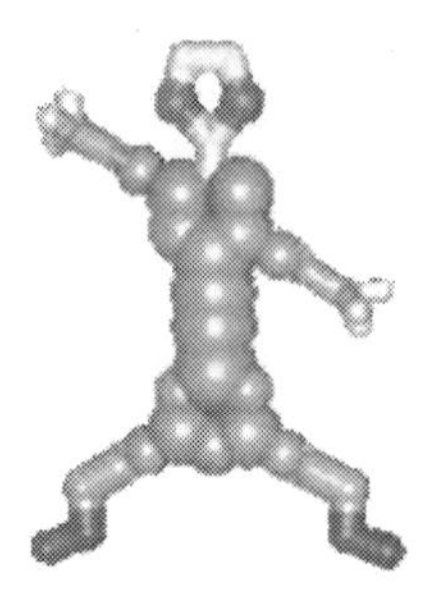

Page Above: the *NanoPutian* family.
Above: *NanoKid* as a spacefill structure.
(Figures reprinted in part with permission from ref.[110]. Copyright 2003 American Chemical Society)

NanoBalletDancer.

Nanocars and Nanotrikes

These are the perfect vehicles for driving *Nanoputians* around their nanoworld. They are made from a rigid framework of benzenes and acetylene groups, with either three or four C_{60} molecules attached at the ends as 'wheels'. I always wanted a compact....

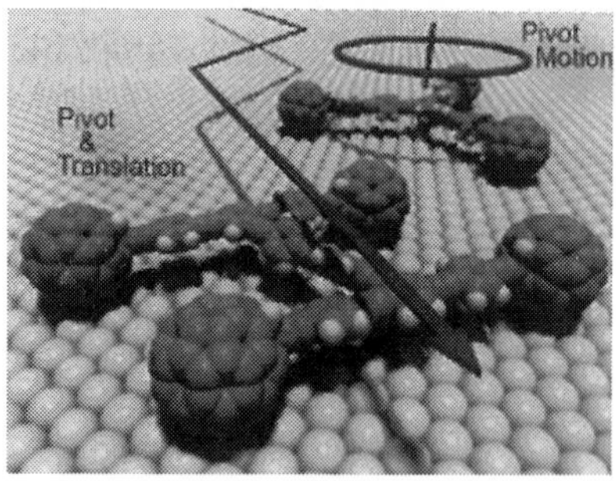

Nano-vroom, nano-vroom!
(Figure reprinted in part with permission from ref.[111]. Copyright 2005 American Chemical Society)

Bicyclohexyl

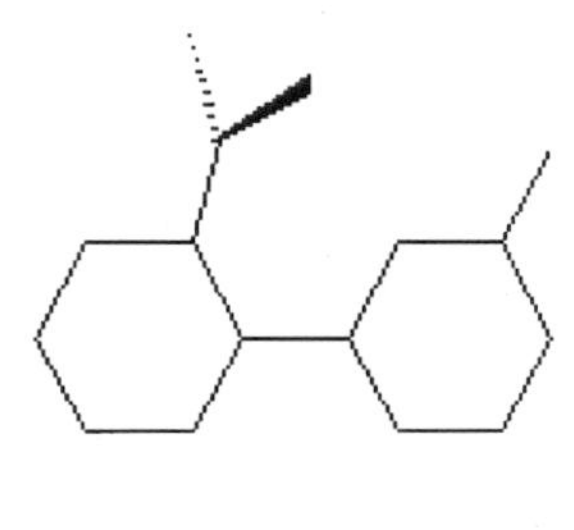

This molecule not only has a name that sounds like a bicycle, it even looks a bit like one too! In fact, the *bicyclohexyl* compound with *isopropyl* and *methyl* sidechains (2-*isopropyl*-3'-*methylbicyclohexyl*, shown in the diagram) looks even more like a bicycle. There is also *tricyclene*, but unfortunately its structure looks nothing like a tricycle.

Hardwickiic acid

This is a *diterpene* which got its name since it was first isolated from the Indian tree *Hardwickia pinnata*[112]. I assume the tree was named after Thomas Hardwicke the English naturalist who lived in India around 1800[113].

Kojic Acid

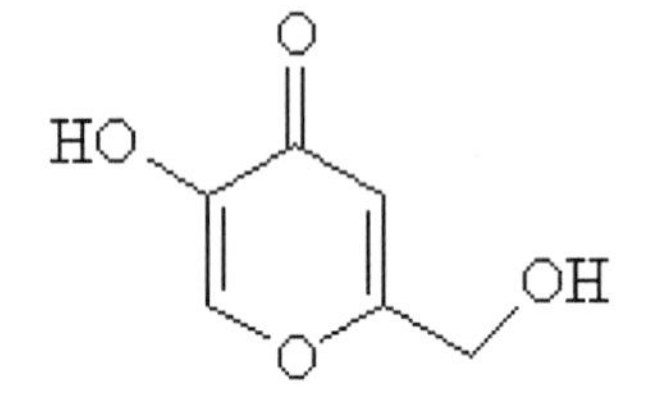

Who loves ya, baby? Tele Savalas as lollipop-loving cop Theo Kojak.
(Copyright Getty Images, used with permission).

This sounds like Tele Savalas' favourite molecule. In case you didn't know, he was the star of the US TV cop show from the 1970s called 'Kojak'. Its official name is 5-*hydroxy-2-(hydroxymethyl)-4-pyrone*) and it is produced by several types of fungi, including *Aspergillus oryzae*, which is called *koji* in Japanese [114]. *Kojic acid* is a by-product in the fermentation process of malting rice, for use in the manufacturing of sake. It is used as a skin whitener... but does it work especially well on bald heads?

CAMP

This effeminate sounding molecule is actually short for 3'-5'-*cyclic adenosine monophosphate* (*cAMP*) and is a signalling molecule which can be found in almost all eukaryotic organisms [115]. For example, it is used as a nutrient sensor in yeast, and is one of the building blocks of DNA. I wonder if it's responsible for the so-called 'gay gene'?

Anisole

Anisole sounds like a molecule the devil would be very interested in collecting, or maybe it's James Brown's ('The Godfather of Soul') favourite molecule? Anisole is aromatic in both the chemical and olfactory sense, and is used in perfumery. It is also an insect pheromone. If you have a lisp, please don't confuse *anisole* with *anethole*, which is a structurally related natural product which, incidentally, has the flavour of aniseed.

James Brown the Godfather of (ani-)sole?
(Copyright Getty Images, used with permission, photographer: Michael Ochs Archives/Stringer).

Aristolochic Acid

This aristocratic sounding molecule is derived from species of the birthwort plant (*aristolochia*) [116]. Plants containing the compound were used in herbal medicine as anti-inflammatory agents and for weight-loss, but they are now banned in the US and Europe as the compound is carcinogenic and toxic to the kidney [117].

Spiroagnosterol

Spiro Agnew in 1972 [118].

I'm told [119] that years ago this molecule was called the "Vice Presidential Steroid" because of the similarity in name to Spiro Agnew, the Vice President of the U.S. from January 1969 until October 1973.

HO

Nootkatone

O

This molecule with a very silly name is used as a food additive to give grapefruit flavours, as well as in the perfume industry to give odours of citrus fruits and orange peel.

It got its name from the yellow cedar, or *Camaecyparis nootkatensis*, which was itself named after the native North American tribe called the *Nootka*.

Sobrerol

I said *sobrerol*, not sombrero!
(Photo: PWM).

HO CH3 CH3
HO
CH3

This isn't the Mexican hat, but a *mucolytic* molecule which dissolves thick mucus usually to help relieve respiratory difficulties. It's named after the Italian chemist Ascanio Sobrero who first determined its structure.

Ladderane

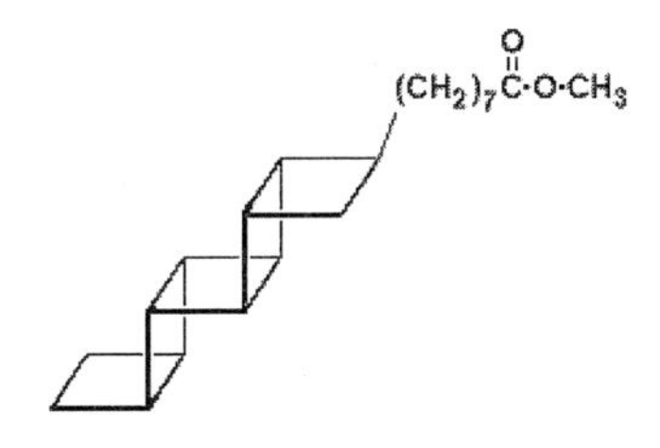

Ladderane is one example of a family of molecules which have a chain of fused *cyclobutane* rings, and which make up the bulk of dense membranes in certain unusual bacteria [120].

They were discovered in *anammox* bacteria, which anaerobically oxidize NH_3 to N_2. The staircase-like structure of *cis*-fused *cyclobutanes* has never before been seen in nature. The most abundant lipid in the bacteria is the *methyl ester* of a C_{20} fatty acid with five fused rings. Other *ladderane lipids* contain three fused *cyclobutane* rings attached to a *cyclohexane*.

Cucurbituril

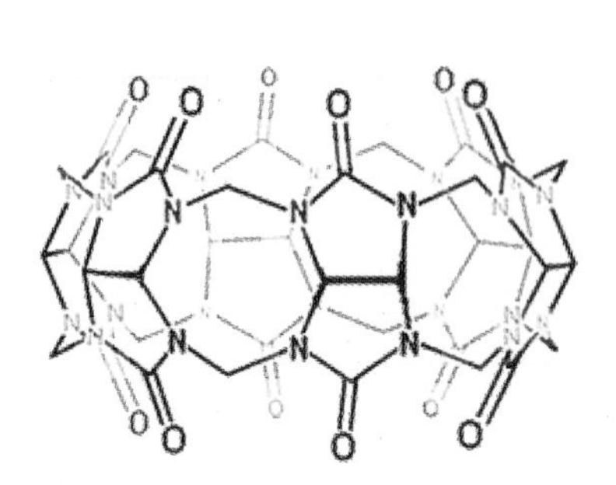

(Copyright Getty Images, used with permission. Photographer: Jose Luis Pelaez).

This molecule, which is shaped like a Halloween Jack-o'Lantern, is named after the Latin word for pumpkin (*Cucurbita pepo*). It is now finding lots of use in medical drugs or in potential molecular electronic devices [121] due to the fact that other long, thin molecules can be threaded through the hole in the centre to make so-called *rotaxanes*.

Mucic Acid

Pronounced '*music acid*', this is quite different to Acid Music... This chemical is obtained by the *nitric acid* oxidation of milk sugar (*lactose*), *dulcite*, *galactose*, *quercite* and most varieties of gum.

It is also called *galactaric acid*. The "*mucic acid test*" in basic bio-chemistry labs is a well-known test for *D-* or *L-galactose*. The test is carried out by oxidising the sample with concentrated *nitric acid*; *mucic acid* crystals will form after leaving the solution overnight. Isn't chemistry great? Just add some acid, and you get some music...

Hipposudoric Acid

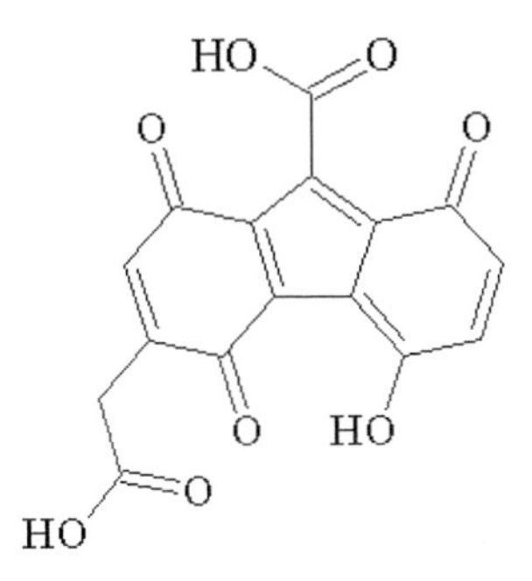

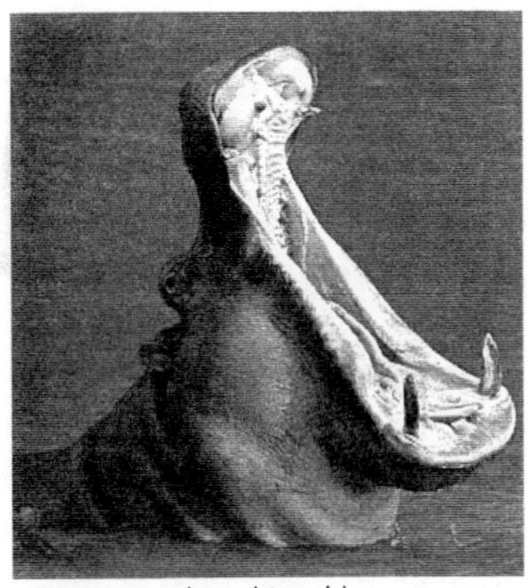

Open up and say 'Aaaah'...
(Copyright Getty Images, used with permission. Photographer: Adam Jones).

Hipposudoric acid is a malodorous blood-red pigment found in hippo sweat [122], and gets its name from hippo + *sudor* (Latin for sweat), so it's literally hippo-sweat acid! It absorbs ultraviolet light, thus blocking out the sun's rays like a sunscreen. It is also a natural antiseptic. Its red colour is responsible for the myth that hippos sweat blood.

And still on the hippo theme (although not the animal this time), there's a molecule called *Hipposulfate A*, which is a poisonous sulfated *sesterterpenoid* that's found in a sponge, *Hippospongia metachromans*, living near Okinawa [123].

NUN

This molecule could be habit-forming... It's actually a linear molecule of *uranium nitride* N-U-N, made by inserting uranium atoms into molecular N_2 [124].

Bongkrekic acid

Any visitor to an Indonesian market or dinner table will almost certainly come across *tempe.* Closely resembling a Camembert cheese in colour and texture with a mushroom-like aroma, *tempe* is in fact one of the world's first soybean foods.

Soya *tempe* in a market in Jakarta.
(Photo: Sakurai Midori[125]).

It is normally harmless, but deep in the mountain villages in Central Java there used to be one rare deadly variety called *tempe Bongkrèk.* This *tempe* is so dangerous that the government has banned manufacture of it and imposed a prison sentence to anybody caught making or selling it, since it can contain a toxin more deadly than cyanide. Made from coconut residue after the oil has been extracted, and unlike good safe soybean *tempe, tempe Bongkrèk* may become contaminated with a bacterium that lives on the fermented coconut, called *Pseudomonas,* which produces the deadly respiratory toxin *Bongkrekic acid*[126]. In 1988 one batch killed 40 people within two days and over a hundred others were hospitalised. Despite the risks, the community in this area have nevertheless continued to eat *tempe Bongkrèk,* so irresistible is the taste and texture of this dangerous and illicit pleasure.

Coproverdine

This molecule is a new anti-tumor drug that was isolated from a sponge which was discovered off the coast of New Zealand. In the National Institute for Water and Atmospheric research (NIWA) archives, it is recorded as being "Green-sheep-sh*t like in appearance".

The alkaloid they discovered was cytotoxic, but they needed a catchy name for it. They settled on *coproverdine*: *copro* = dung, *ovis* = sheep, and *verdi* = green [127].

SnOT

Tritiated tin hydroxide goes by the wonderful chemical formula of *SnOT*. David Ball, a chemist in Cleveland Ohio, was working on isotopomers of SnOH, and after tritiating it found he got SnOT. His paper [128] concludes with the wonderful phrase "Since Wang *et al* did not use tritium substitution, we can state with certainty that there was no SnOT in their samples".

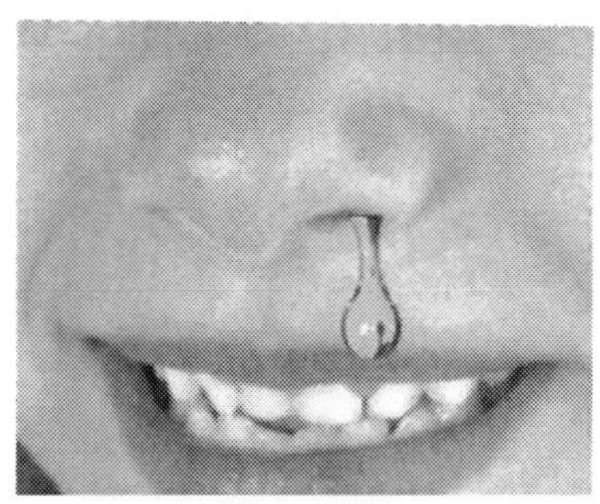

(Photo: PWM)

Bullvalene

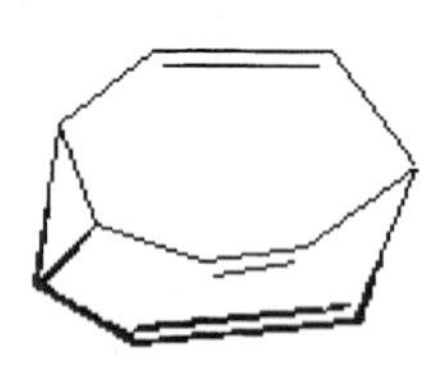

This is a very unusual molecule, in that it is fluxional...all the carbons are equivalent due to the rapid movement of the double bonds around the structure [129]. It was first predicted to be like this over 20 years ago by Professor 'Bull' Doering, and was only synthesised in the lab many years later, whereupon his controversial predictions about the structure were verified.

Is this the correct structure - or just a load of 'bull'?

The name is thought to be derived from his nickname, 'Bull', but other reports [5] suggest that it was given its name by an irreverent and skeptical graduate student who thought such a structure couldn't exist, and was just a load of 'bull'.

Barrelene

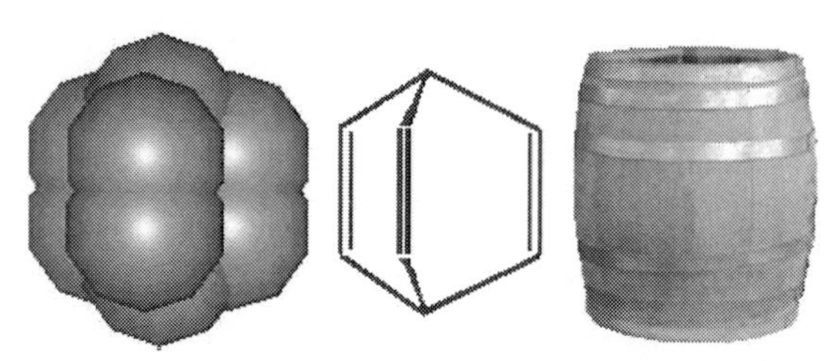

This molecule is closely related to *bullvalene*, and got its name from its similarity to the shape and structure of a barrel, surprisingly enough [130].

Chavicine

Caricature of a British 'chav'!
(Drawn by J.J. McCullough [131])

For non-UK readers, 'chav' is a British slang term for a subcultural stereotype of a youth who's fixated on fashions such as imitation gold, poorly made jewellery and fake designer clothing, combined with elements of working class British street fashion, such as trainers, tracksuit bottoms and polo shirts. *Chavicine* sounds like an ideal molecule for them, since it's the hot, sharp flavour found in black pepper - perfect to put on a late night pizza or kebab on the way home from the pub.

Flufenamic Acid

This molecule with a very fluffy name has anti-inflammatory and antipyretic properties, and is used to treat inflammatory rheumatoid diseases and relieve acute pain.

Its chemical name is *N-*(3-*trifluoromethylphenyl*)*anthranilic acid*, which doesn't sound quite as friendly and fluffy as its common name. In other languages it is called *Flufenaminsäure* (German), *flufenaminezuur* (Dutch), ácido *flufenamico*! (Spanish), and *acide flufenamique* (French).

DOPE

'DOPE'
1,2-Dioleoyl-sn-Glycero-3-Phosphoethanolamine

Apparently, *DOPE* is commonly used by membrane chemists and biochemists - which is something I've always suspected... It's actually short for 1,2-*Dioleoyl-sn-Glycero*-3-*Phosphoethanolamine*, and it's a *phospholipid* used for research into membrane structures. A variant on this is called *DOGS*, so if you take *DOPE*, you may go to the *DOGS*[132].

SNOG

Left: Britney Spears and Madonna sharing a *S-Nitrosoglutathione* at the 2003 MTCV awards.

(Copyright Getty Images, used with permission. Photographer: Chris Polk).

SNOG is a utility carrier of *nitric oxide* which breaks down to produce *nitric oxide* and a *glutathione* radical at pH 7.4. Its proper chemical name is *S-Nitrosoglutathione*. Unsurprisingly, one effect of *SNOG* is that it apparently causes smooth muscle relaxation... For those of you unfamiliar with UK slang, a 'snog' is a deep passionate kiss, similar to the one Britney is giving Madonna in the photo above.

Hirsutene

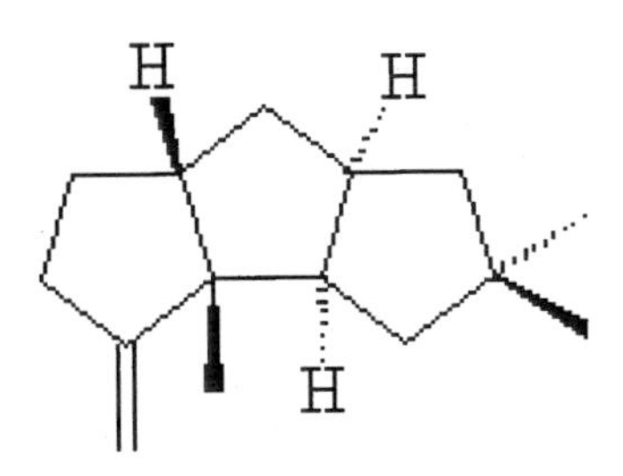

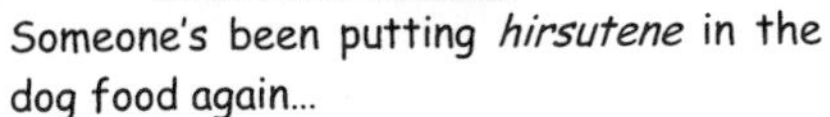
Someone's been putting *hirsutene* in the dog food again...

This is an important antibiotic that's derived from various fungi that live on dead wood, including *Stereum hirsutum* which provided its name [133]. Perhaps a side-effect of this molecule is to makes things hairy..? Or maybe it counteracts the effects of *boldine*, mentioned earlier...

Wrenchnolol

This is a molecule that looks like a monkey wrench, and is an anti-cancer drug [134]. The "jaw" part of the compound mimics part of a transcription factor molecule, and the "handle" region accepts chemical modifications for a range of analysis.

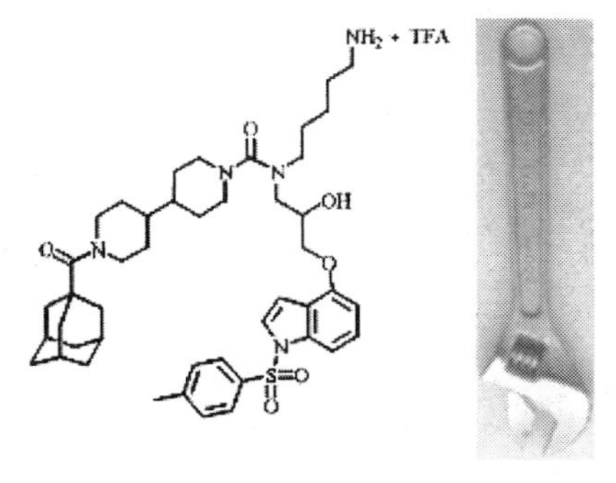

Forskolin

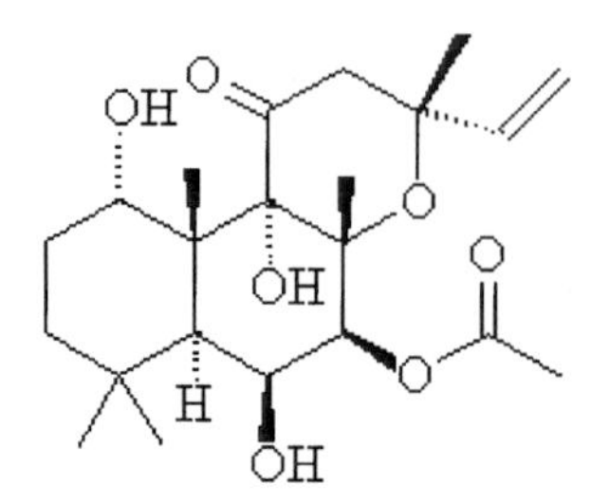

Despite the odd name, *forskolin* is not what they remove from the baby during ritual circumcision. In India, practitioners of traditional *Ayurvedic* medicine have long used the herb *Coleus forskohlii* to treat asthma, heart disease, and a range of other ailments.

In the 1970s, researchers isolated a chemically active ingredient in the herb and called it *forskolin*[135]. Now available in supplement form, this extract is commonly recommended for treating hypothyroidism, a condition in which the thyroid gland produces too little thyroid hormone. *Forskolin* is believed to stimulate the release of this hormone, thus relieving symptoms as fatigue, depression, weight gain, and...of course... dry skin.

Indenyl

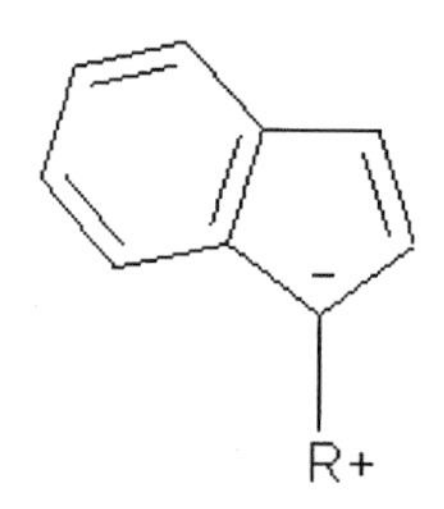

This sounds like it could be a pollutant found in some Egyptian rivers (*i.e.* in de Nile...), but it's actually a fusion of a *cyclopentadienyl* ring with a *benzene* ring, and is often used as a ligand for *metallocene* synthesis.

Ru(Tris)BiPy-on-a-stick

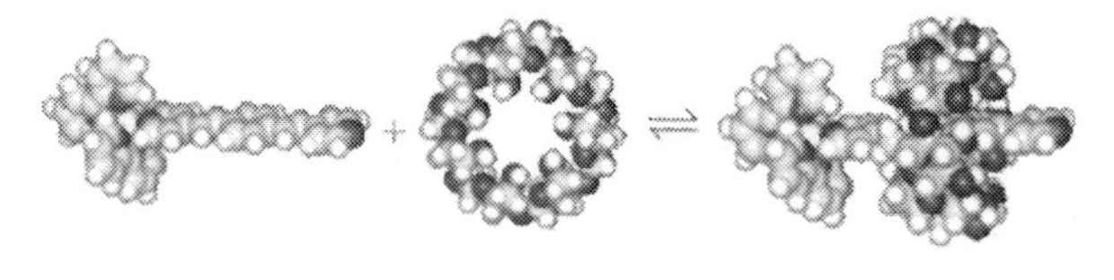

Ru(Tris)BiPy-on-a-stick (left) plus *cyclodextrin* (right) gettin' it together [136].
(Figure reproduced with permission of the American Chemical Society).

This one is a bit of a cheat, since this name (pronounced Rew-Tris-Bip-ee on a stick) is not officially recognised. But I'm reliably informed that inorganic chemists who deal with this compound use this nickname regularly to avoid having to deal with its even more cumbersome IUPAC name. The other reason to include it is the wonderfully suggestive figure from the paper (above) of the molecule threading through the hole in *cyclodextrin*.

Skunky Thiol

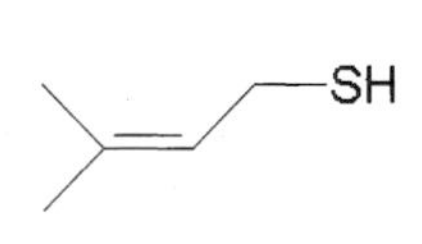

This molecule is what makes beer taste bad after it's been left exposed to sunlight for a few hours [137].

The actual name is 3-*methylbut*-2-*ene*-1-*thiol*, or 3-*MBT* for short, but since it's related to molecules found in skunk spray, and it stinks, it's also known as *skunky thiol*. Only a few nanograms of this *thiol* in one litre of beer are enough to give the offensive flavour.

Banana Borane

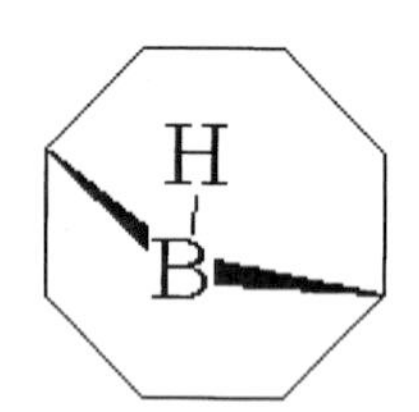

This isn't an official name, but I'm told that many chemists who work in the *organoborane* field use the nickname *banana borane* to describe molecules such as 9-*borabicyclo*[3,3,1]*nonane*, abbreviated *BBN*. This is because rather than draw out the proper structure (top), they simply draw the *borane* as a banana shape with the bridging B group sticking out (bottom).

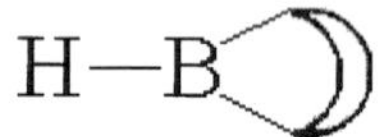

Thebacon

The bacon and the eggs and the coffee.

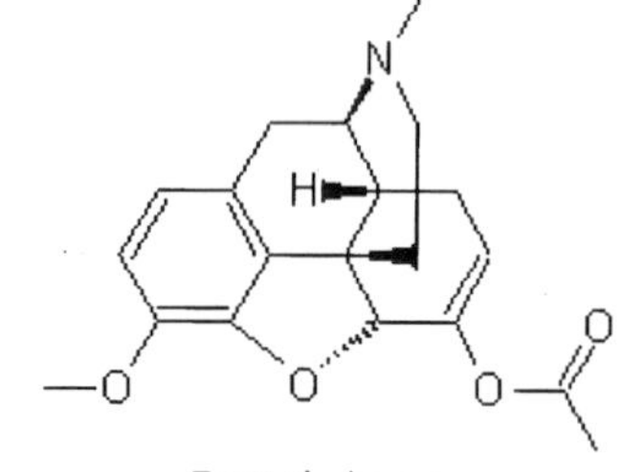

Just thebacon.

This molecule certainly brings home the-bacon! It has a similar structure to *diamorphine* (*heroin*) but has only one *acetyl* group instead of two, and the other group is replaced with a CH_3O- group. Apparently, *thebacon hydrochloride* is a centrally acting cough suppressant sometimes used to treat coughs. Maybe it should taken together with *sandwicensin* to get 'the bacon sandwich'. *Thebacon* is derived from *thebaine* (would doing the synthesis be *the bane* of some poor chemist's life!) whose name comes from the Latin word *thebacus*, meaning 'from Thebes'. *Thebain* is found in *opium*, and since in the 19th century much of the world's *opium* came from Egypt, it was named after its origin, the city of Thebes.

Cryogenine

Cryogenine A

Cryogenine B

Cryogenine A (also known as *vertine*) is the active constituent of the shrubby yellowcrest (*Sinicuichi*) plant [138].

It can give audio hallucinations and produces mild euphoria and aching muscles in large doses. I don't know why it is called *cryogenine...* maybe because after you stop using it you go through very, very cold turkey? In fact there are two totally unrelated, and different molecules called *cryogenine.* The other molecule, *cryogenine B,* is usually called 1-*phenylsemicarbazide,* is used to reduce fever - which is appropriate given its name!

Damsyl

When this molecular fragment reacts, could it be called a '*damsyl* in distress'? The name is coined by contraction of its full name, 4-(*dimethylamino*)*benzenesulfonyl.*

Lagerine and Bebeerine

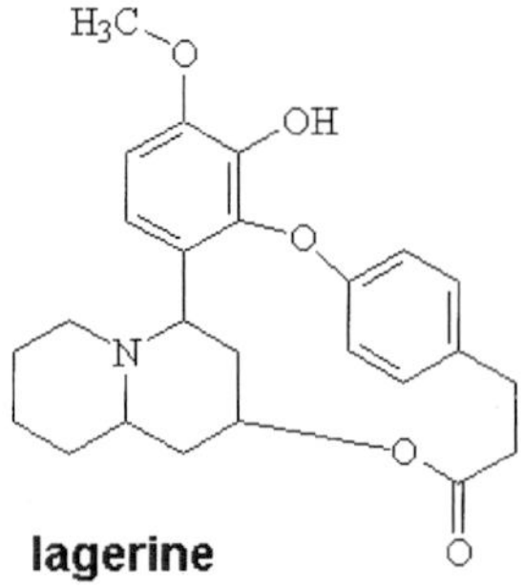

lagerine

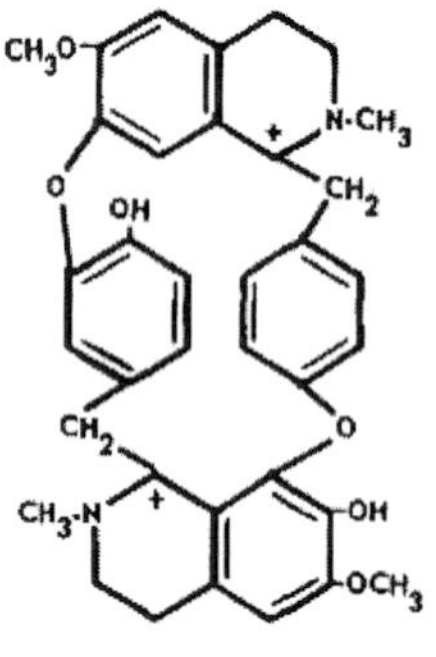

bebeerine

I wonder if *lagerine* is sold by the pint? It actually has nothing to do with beer, it gets its name from being a constituent of the crape myrtle (*Lagerstroemia indica L.*) plant [139]. *Bebeerine* too, has nothing to do with beer. It's an alkaloid molecule derived from the Caribbean *bebeeru* tree, and helps to protect it from attack by beetles [140].

PORN

Now that I've put the word *PORN* in this book, it'll either get banned or increase the sales hugely! Unfortunately, this *PORN* is simply the acronym for *poly-L-ornithine*, a molecule used in cell culture experiments. I suppose that forming a polymer does involve lots of frantic couplings...so *PORN* maybe isn't such an inappropriate name after all.

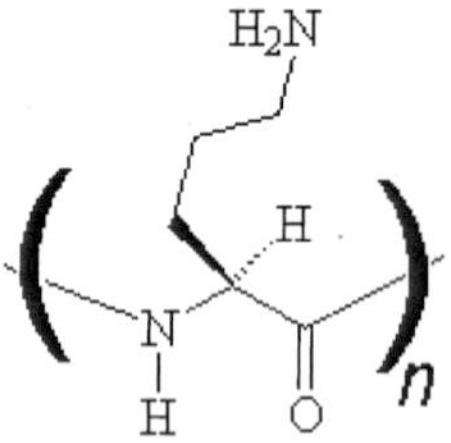

Jawsamycin, Histrionicotoxin and Yessotoxin

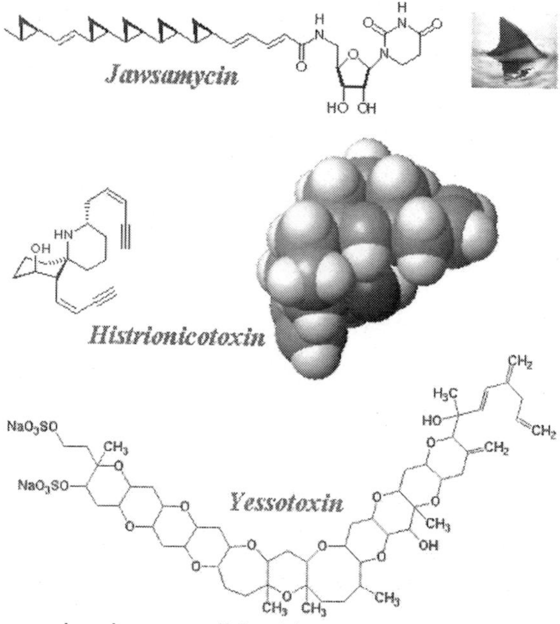

These three molecules are all highly toxic, and are normally isolated from biological sources. *Jawsamycin* was discovered in 1990 by Fujisawa [141], a Japanese pharmaceutical company, but it has only recently been synthesised. The metabolite is composed of a chain of five triangular *cyclopropyls*, which giving it its name because of the resemblance to shark's teeth [142,143]. The compound, however, has 10 chiral centres - for a total of 1024 possible isomers, only one of which is *jawsamycin*. It can be used to fight fungal infections - the molecule just gobbles them up!

Histrionicotoxin sounds dramatic - and it is! It's a poison found on the skin of a tree frog (*Dendrobates histrionicus*) in South America, and is used by the native Indians on their blow-pipe darts [144].

Yessotoxin was first isolated from the digestive organs from scallops (*Patinopecten yessoensis*) in Japan and is believed to be produced by microalgae [145].

Sarcosine

This is one for French readers. Nicolas Sarkozy is currently the President of France, and, as such, it's appropriate that he has his own molecule. *Sarcosine* is a sweetish crystalline *amino acid* found in muscles and other tissues, and is also called *N-methylglycine*. It was discovered and named by Liebig (presumably using his condenser), and is present in foods such as egg yolks, turkey, ham and vegetables. It's also used in the production of toothpaste.

Nicolas Sarkozy
(Copyright Getty Images, used with permission, photographer: Nicholas Roberts/AFP).

Another molecule which sounds odd in French (but not in English) is *pyralene*, which has been used in the past as an insulating oil in electric transformers. In French it is pronounced *pire haleine*, which means "worst breath". This is ironic, since use of *pyralene* was abandoned after transformer fires were giving off toxic fumes.

Enflurane

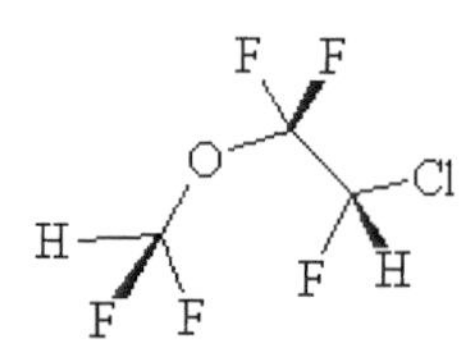

Yet another one for the French speakers... *Enflure* in French means a twit, a clot, or a jerk, or can also mean a swelling or inflammation. *Enflurane* is an outdated *halogenated ether* that was used for inhalation anesthesia in the 1970's and 80's.

Shikimic Acid

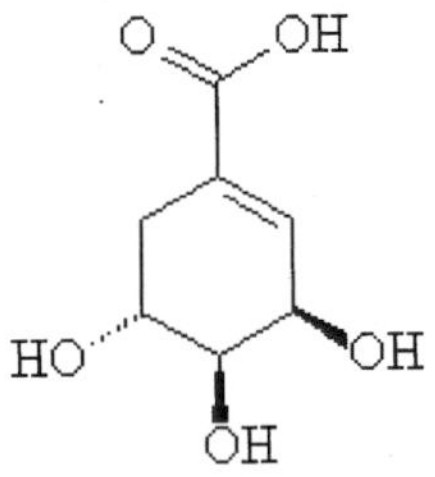

In the interests of fairness, I should include a molecule for German-speaking readers, too. *Shikimic* sounds very like the German word *Schickimicki*, which means a (snobbish) member of the in-crowd, a trend-setter, or a 'designer-label-wearing' sort of person. But the trendy *shikimic acid* was first isolated from the Japanese flower *shikimi*, hence its name. It's used as a starting molecule in the synthesis of the anti-flu drug *Tamiflu*[146].

Naftazone

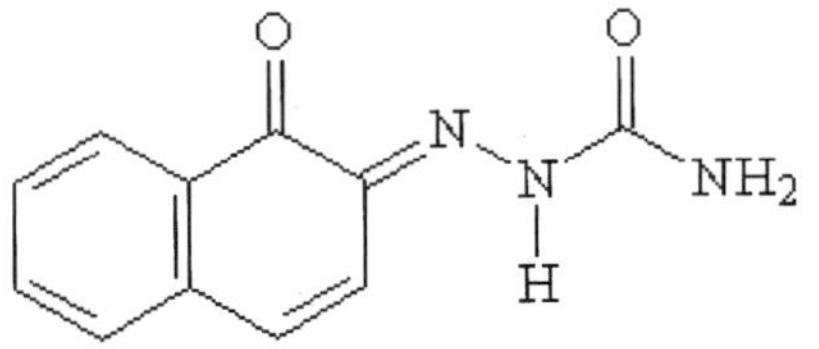

This sounds like a pretty naff molecule. (For non-UK readers, 'naff' is English slang for something that's poor quality, unfashionable or just plain rubbish).

It's used as a drug to protect blood vessels (a 'vasoprotector') - so, maybe it's not quite so naff after all. Its name comes from a contraction of its official name (1,2-*naphthoquinone*-2-*semicarbazone*) and it has nothing whatsoever to do with the North American Free Trade Agreement (NAFTA).

Mandelic Acid

Mandelic acid is not named after Nelson Mandela, the world famous South African politician and winner of the 1993 Nobel Peace Prize, although his youthful appearance might be due to it....as *mandelic acid* is often used in skin creams to smooth away wrinkles. It's also used as an antiseptic ingredient particularly against urinary tract infections.

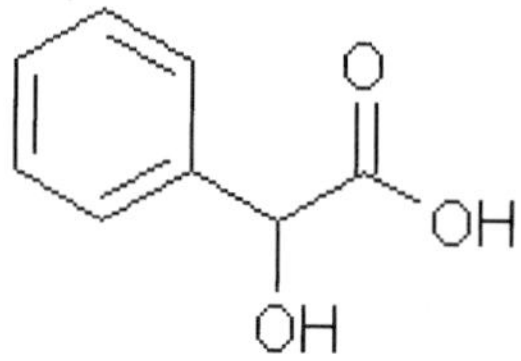

Nelson Mandela in 2007
(Copyright Getty Images, used with permission, photographer: Gianluigi Guercia/AFP).

It's named after the German word *Mandel*, meaning almonds, since it's extracted by heating bitter almonds with HCl.

Olympiadane

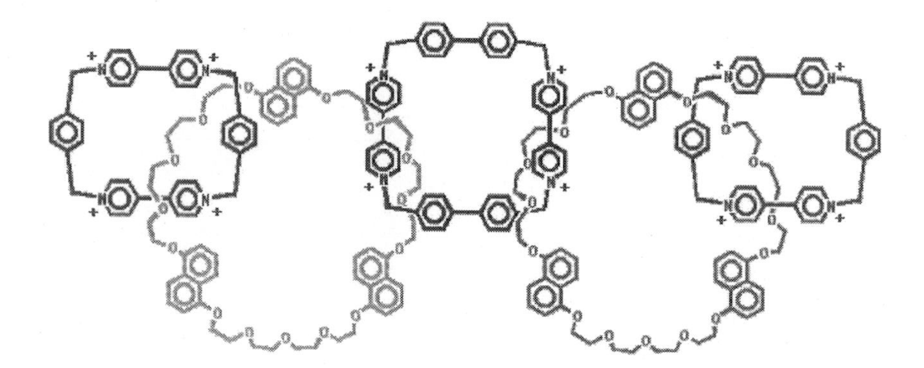

This molecule consists of five interlocking rings, which mimics the Olympic Games symbol, and so is named *olympiadane*[147]. It was first made in 1994, in commemoration of the Olympic Games due in 2 year's time. The successful linkage of these highly complex synthetic molecules means that molecular chains of any length could be constructed with many applications, particularly in the areas of information storage systems and the creation of a 'molecular computer'.

Puberulin

This wonderfully named molecule gets its name from the fact that it's isolated from the African shrub *A. puberula.* Hmm, I wonder what you'd measure with a *pube-rula...?*

Fartox

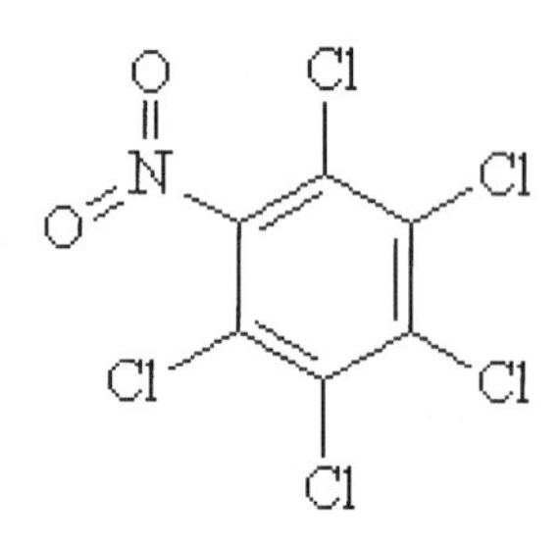

Guess who's been using *Fartox*...
(Copyright Getty Images, reproduced with permission Photographer: Digital Vision).

On a related theme, this molecule is actually called 1,2,3,4,5-*pentachloro*-6-*nitrobenzene*, but goes by a variety of rather silly tradenames [148], including *Quintozene, Folidol, Tubergran, Terrafun, Earthcide* and best of all, *Fartox*. It seems somehow appropriate that *Fartox* should be a pale yellow solid with a slightly musty odour. It has been used as a soil fungicide since the 1930's, but I have no idea how it came to have this silly name. Maybe it was due to a side-effect of eating fruit sprayed with it?

Biline

A newspaper reporter's favourite molecule? (They always want their by-line). This is a bile pigment, and there are various versions depending where the H goes (in the diagram it's on N-21, so the molecule shown is 21*H-biline*), and what side chains (if any) there are.

Dinocap

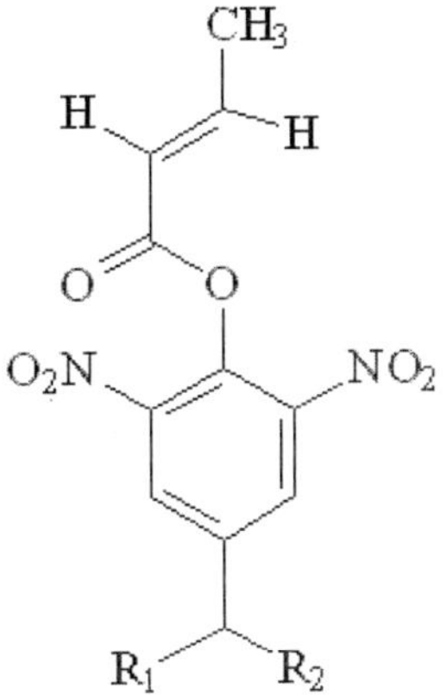

Dinocap sounds like a dinosaur's hat, or a hat that looks like a dinosaur. It's a dark red viscous liquid that's used to kill mites, fungus and mildew on crops, and goes by the trade name *Karathane.* There are 2 different versions of *dinocap*; *dinocap*-4 is the one shown on the right and *dinocap*-6 swaps the positions of one of the NO_2 groups and the $R_1(CH_2)R_2$ side-group. In either case, the liquid is actually a mixture of different isomers, with $R_1=CH_3(CH_2)_n$ and $R_2=CH_3(CH_2)_{5-n}$.

I'm pretty sure the '*din-*' part of the name comes from *dinitro-*, the '*-oc-*' from *octylphenyl crotonate*, but the '*-ap*' remains a mystery.

Tamuic Acid

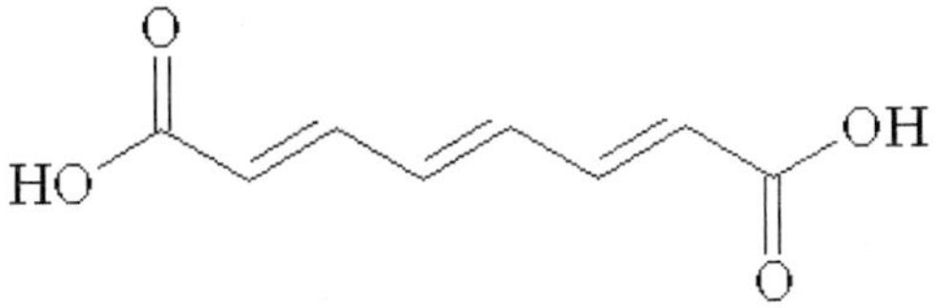

Tamuic acid is a shorthand term for an organic acid with a huge unwieldy name that's used as a ligand in organometallic chemistry.

The name comes from the place it was discovered, Texas A&M University [149]. The related molecule with 4 C=C units was named by the same group *texic acid* (after Texas). On the same theme, there's also *tuftsin*, a *tetrapeptide* named after Tufts University in Boston where it was discovered. If the trend continues, let's hope that Cambridge University Nano-Technology centre don't discover a molecule soon!

Discodermolide

Could this be John Travolta's favourite molecule? It's a recently discovered *polyketide* natural product found to be a potent inhibitor of tumor cell growth, and it gets its name since it was first isolated in 1990 from the Caribbean marine sponge *Discodermia dissoluta*[150]. Since the compound is light-sensitive, the sponge must be harvested at a minimum depth of 33 metres - so, no disco lighting there then...

Right: Does swallowing *discodermalide* bring on a Saturday night fever?

(Photo: John Travolta in 'Saturday Night Fever', copyright Getty Images, reproduced with permission. Photographer: Michael Ochs archives).

Pregnane

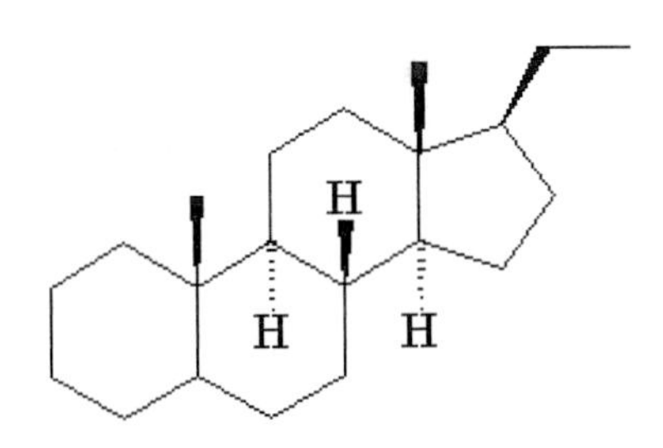

As you might expect, *pregnane* is the 'parent' steroid for many hormones, including the one important during pregnancy, *progesterone.*

She's been at the *pregnane* again...
(Copyright Getty Images, reproduced with permission. Photographer: Susanne Walstrom).

Asparagine

I wonder if this molecule tastes of asparagus? In fact, it gets its name since it was first isolated in 1806 from asparagus juice, and was the first *amino acid* to be isolated. It is one of the 20 most common natural *amino acids* on Earth and can be synthesised in the body.

(Photo: PWM)

Interestingly, the smell observed in the urine of some individuals after consumption of asparagus is attributed to a byproduct of the metabolic breakdown of *asparagines* called *asparagine-amino-succinic-acid monoamide.* However, some scientists disagree and implicate other substances in the smell, especially *methanethiol.*

Gingerol

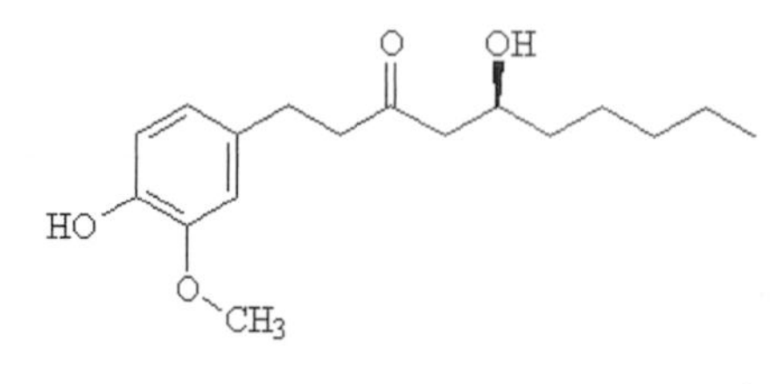

Right: Was Elizabeth I a fan of *gingerol*[151].

Gingerol isn't the molecule responsible for ginger hair and freckles, or even gingivitis. Instead, it's the active constituent of fresh ginger. *Gingerol* is a relative of *capsaicin*, the compound that gives chilli peppers their spiciness. It's normally found as a pungent yellow oil, but also can form a low-melting crystalline solid. Cooking ginger transforms *gingerol* into the compound with a 'zing', *zingerone*.

Bionic Acids

I wonder if these cost six million dollars and give you super strength? Actually, there are a number of *bionic acids* which are derived from *cellobiose*, which is a sugar digestion product of *cellulose*.

The differences come from the component sugars which make up the reduced form, and include *maltobionic*, *melibionic*, *cellobionic*, *aldobionic* and *lactobionic acid* (the one shown above).

Porphyrin Hamburger

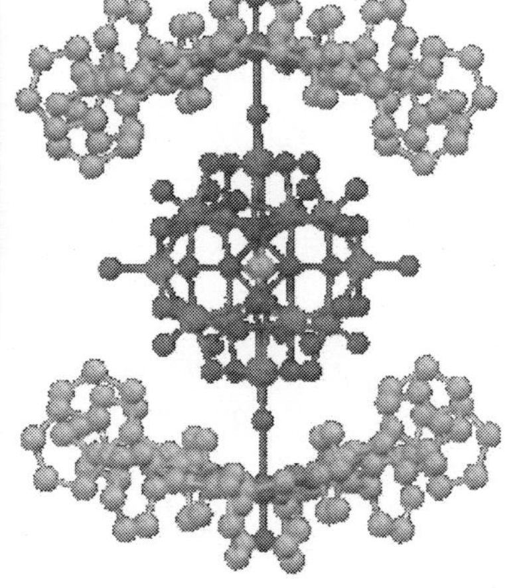

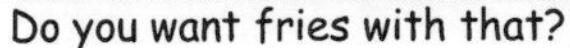

Do you want fries with that?

Above: A real hamburger [152].

Right: A *porphyrin hamburger*
(Image reproduced with permission of the RSC from ref.[153]).

A team of researchers at Osaka University, Japan, fused a *molybdenum-porphyrin* complex and a *tungsten polyoxometalate* to form a compound they have named the *porphyrin hamburger* [153]. Two saddle-shaped *porphyrin* complexes make up the burger buns, while a cluster of *tungsten oxide* anions surrounding a central silicon cation, known as a *polyoxometalate*, forms the meat sandwiched between them. The molecules are joined by stable coordination bonds.

This must give plenty of opportunity for derivatives along the lines of *porphyrin cheeseburger*. Or perhaps you could even attach *penguinone* to make a *penguin burger*? Or even attach a molecule of *cocaine* to get a burger and coke?

Blasticidin

And here's one for the military, or maybe just a gung-ho chemist (Blast its side in!). It's actually an antibiotic fungicide, and is also used in genetic engineering experiments to fuse pieces of DNA together to make resistant genes.

Bender's salt and Wanklyn's soap

Is *Bender's salt* what gay chemists put on their fries? Or maybe it's the salt that Bender from the TV cartoon series 'Futurama' would use? *Bender's salt* (*potassium ethylthiocarbonate*, $C_3H_5KO_2S$) is named after the German chemist Friedrich Bender.

On a similar theme, there's *Wanklyn's soap* (used a lot by teenaged boys, no doubt!). According to the label it's flammable and has one degree of hardness! *Wanklyn's Soap* is an ethanolic solution of soap which was formerly used to test for water hardness, and formed the basis of the Wanklyn Scale of hardness.

CuNT

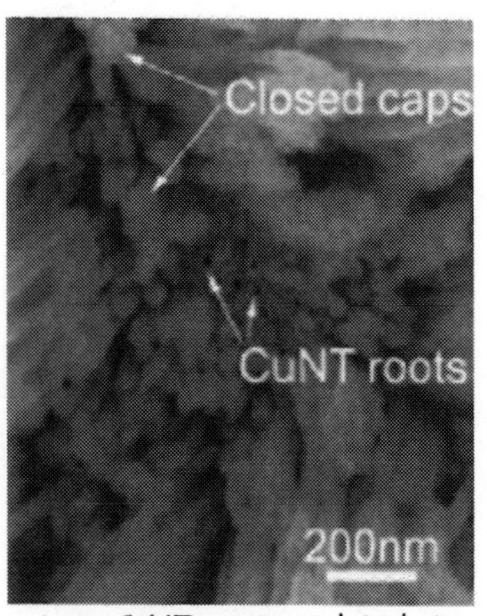

Are *CuNT* roots related to beetroots?
(Image used from ref.[154] with permission of the RSC).

This one must be rudest of all acronyms! *Carbon nanotubes* are often abbreviated to *CNTs*, and single-walled ones to *SWCNTs*. But, unbelievably, when a Chinese group recently fabricated *copper nanotubes*[154], they decided to call them *CuNTs*! In the same paper they describe *bismuth nanotubes*, and called them *BiNTs*.

Either they named these 2 structures for a bet - just to see if the Royal Society of Chemistry would publish a paper containing numerous (over 50!) references to *BiNTs* and *CuNTs* - or they just didn't realise the meanings of these two acronyms. Or maybe they just did it to increase the number of hits they receive from online searches.

Folk Acid

This is nothing to do with folk music, or even acid folk, it is simply a mis-spelling of *folic acid*, which itself gets its name from the Latin word *folium* meaning 'leaf'. This seems to be a particularly common mis-spelling, and occurs even in scientific papers and textbooks (try searching for it in Google), maybe a result of word-processing programs automatically 'correcting' words they don't recognise, or just that scientists can't spell.

Scorpionate ligands

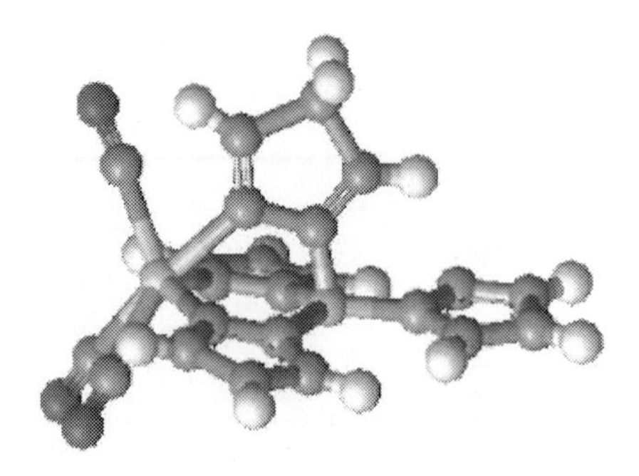

The *scorpionate ligand* gets its name from the fact that it can bind to a metal with three donor sites. The first two are like the pincers of a scorpion [155].

The third donor site (the 'tail') reaches over the plane formed by the metal and the 'pincers' to bind to the metal from above, like a scorpion grabbing the metal with two pincers before stinging it. The most popular class of *scorpionates* are the *Tp ligands* (*tris(pyrazolyl)hydroborates*), and, above, one is shown bonding to a $Mn(CO)_3$ group. Another scorpion-like molecule is *bis* ([1,2]*dithiolo*)-[1,4]*thiazine*, which is sometimes unofficially called *sscorpionine* [156].

Penicillin

I thought I'd end this section with the most well-known antibiotic compound. Although everyone has heard of *penicillin*, and some may even know it's named after the *Penicillium* mould, very few realise that this came originally from the Latin word for brush (*penicillus*), and that this ultimately came from the diminutive of *penis* (meaning 'tail' or 'male member'). This is quite appropriate, since one of the first uses for this 'small penis' antibiotic was to treat soldiers suffering from VD during WW2!

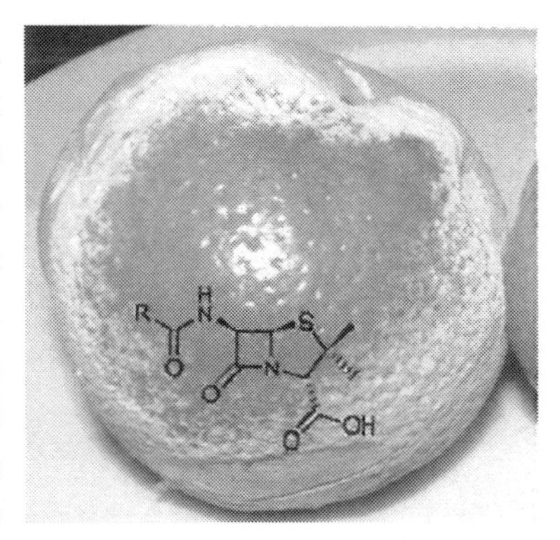

Penicillium mold on a mandarin orange [157], and the core structure of *penicillin* where R is a variable side-group such as *phenyl*.

Molecules named after places

This isn't all of them, but there are quite a few: *himalayamine, pakistanamine, americanin, ecuadorin, grenadadiene, virginiamycin, mauritiamine, alaskene, texaphyrin, alaskaphyrin, taiwanins, montanastatin, mediterranenols, bahamamide, arizonins, pacifenol, brazilin, argentinine, guyanin, jamaicin, louisianins, floridanolide, oregonenes, utahin, michigazone, ukrain, malaysic acid, thailandine, mongolicains, vanuatine, australinols, tasmanine, vietnamine, angolamycin, gabonine, senegalene, madagascarin, tanzanene, ugandoside, yemenimycin, syriamycin, jordanine, atlantone, mexicanolide, panamine, albanols, srilankenyne, seychellogenine* and *borneol*.

References

[1] G. Markl and H. Hauptmann, *J. Organomet. Chem.*, **248** (1983) 269, and also: G. Märkl and H. Hauptmann, "Unusual Substitution in an Arsole Ring", *Angew. Chem.* **84**, (1972) 439.
[2] M. Johansson, *Letts. Org. Chem.* **2** (2005) 469.
[3] A.H. Holm, L. Yusta, P. Carlqvist, T. Brinck, Kim Daasbjerg, *J. Am. Chem. Soc.* **125** (2003) 2148.
[4] http://dictionary.oed.com
[5] A. Nickon and E.F. Silversmith, *Organic Chemistry: The Name Game*, Pergamon, 1987
[6] P.v.R Schleyer, E. Osawa, M.G. B. Drew, *J. Am. Chem. Soc.* **90** (1968) 5034.
[7] S.M. Kupchan, K.L. Stevens, E.A. Rohlfing, B.R. Sickles, A.T. Sneden, R.W. Miller, R.F. Bryan *J. Org. Chem.*, **43** (1978) 586.
[8] See J. Chatt, *Pure and Appl. Chem.* **51** (1979) 381 for the naming scheme.
[9] P.L. Majumdar, R.N. Maity, S.K. Panda, D. Mal, M.S. Raju, E. Wenkert, *J. Org. Chem.* **44** (1979) 2811.
[10] R. Huisgen H. Gotthard, H.O. Bayer, F.C. Shaefer, *Chem. Ber.* **103** (1970) 2611.
[11] A.D.J. Haymet, *J. Am. Chem. Soc.* **108** (1986) 319.
[12] *"Fullerenes"*, R.F. Curl and R.E. Smalley, *Scientific American,* October 1991.
[13] J.C. Hummelen, M. Prato F. Wudl, *J. Am. Chem. Soc.*, **117** (1995) 7003.
[14] R. Kaslauskas, R.O. Lidgard, P.T. Murphy, R.J. Wells, *Tetrahedron Lett.* **21** (1980) 2277.
[15] K. Kohara, R. Kadomoto, H. Kozuka, K. Sakamoto, Y. Hayata, *Food Sci. Technol. Res.* **12** (2006) 38.
[16] T. Ishiyama, M. Murata N. Miyaura, *J. Org. Chem.* **60** (1995) 7508.
[17] A.F. Thomas, B. Willhalm, *Tetrahedron Lett.* **5** (1964) 3177
[18] J.E. Rodgkins, S.D. Brown J.L. Massigill, *Tetrahedron Lett.* **8** (1967) 1321.
[19] G. Savona, S. Passannanti, M.P. Paternostro, F. Piozri, J.R. Hanson, P.B. Hitchcock, Michael Siverns, *J. Chem. Soc. Perkin. Trans. 1* (1978) 356.
[20] M. Konoshima, Y. Ikeshiro, *Tetrahedron Lett.* **11** (1970) 1717.
[21] M. Anchel, *J. Am. Chem. Soc.* **75** (1953) 4621.
[22] http://commons.wikimedia.org/wiki/Image:Michelangelos_David.jpg; GNU free documentation license.
[23] S. Begum, S.B.Usmani, B.S. Siddiqui, S. Siddiqui, *Phytochem.* **36** (1994) 1537.
[24] http://commons.wikimedia.org/wiki/Image:Melon.jpg, public domain
[25] http://commons.wikimedia.org/wiki/Image:FraiseFruitPhoto.jpg;GNU free document license.
[26] E.M. Engler, J.D. Anglose, P.v.R. Schleyer, *J. Am. Chem. Soc.*, **95** (1973) 8005.
[27] K.B. Wiberg, L.K. Olli, N. Golembeski, R.D. Adams, *J. Am. Chem. Soc.* **102**, (1980) 7467.
[28] J.F. Liebman, A. Greenberg, *Chem. Rev.* **76** (1976) 311.
[29] A. Szent-Györgyi "Lost in the Twentieth Century", *Ann. Rev. Biochem.* **32** (1963) 1.
[30] R.A. Klein, G.P. Hazlewood, P. Kemp, R.M.C. Dawson, *Biochem. J.* **183** (1979) 691.
[31] http://commons.wikimedia.org/wiki/Image:Veuve_clicquot_bottle_sizes.jpg, public domain, photo: Walt Nissen.
[32] A.V.R. Rao, K. Venkatar, P. Chakraba, A.K. Sanyal, P.K. Bose, *Indian J. Chem.* **8** (1970) 398.
[33] Image created from the 3D structural file obtained from http://www.pdb.org; structural file PDB ID:1ont, N. Skjaerbaek, K.J. Nielsen, R.J. Lewis, P. Alewood, D.J. Craik, "Determination of the solution structures of conantokin-G and conantokin-T by CD and NMR spectroscopy", *J. Biol. Chem.* **272** (1997) 2291.
[34] J.A. Haack, J. Rivier, T.N. Parks, E.E. Mena, L.J. Cruz, B.M. Olivera, *J Biol. Chem.* **265** (1990) 6025.
[35] J. Altman , E. Babad, J. Itzchaki, D. Ginsburg , *Tetrahedron, Suppl.* **8**, (1966) 279.
[36] P.E. Eaton, T.W. Cole, *J. Am. Chem. Soc.*, **86** (1964) 3157.
[37] H. Taguchi, P., Kanchanapee, T. Amatayakut, *Chem. Pharmaceut. Bull.*, **25** (1977) 1026.
[38] http://www.phytochemie.botanik.univie.ac.at/herbarium/stemona.htm
[39] D.C. Craig, M. N Paddon- Row, *Aust. J. Chem.* **40** (1987) 1951.
[40] N.K. Singh, H.S. Chae, I.H. Hwang, Y.M. Yoo, C.N. Ahn, S.H. Lee, H.J. Lee, H.J. Park, H.Y. Chung, *J. Anim Sci.* **85** (2007) 1126.
[41] A. Nahrstedt, M. Hungeling, F. Petereit, *Fitoterapia* **77** (2006) 484.
[42] D.O. Chester and J.A. Elix, *Austr. J. Chem.* **32** (1979) 2565.
[43] http://en.wikipedia.org/wiki/Image:Pygoscelis_papua.jpg; creative common license.
[44] E. Clar, *Polycyclic Hydrocarbons* (Academic, New York, 1963).
[45] I.I. Creaser, J.MacB. Harrowfield, A.J. Herlt, A.M. Sargeson, J. Springborg, R.J. Gene, M.R. Snow, *J. Am. Chem. Soc.* **99** (1977) 3181.
[46] W.-D. Fessner, H. Prinzbach, G. Rihs, *Tetrahedron Lett.* **24** (1983)5857.
[47] http://commons.wikimedia.org/wiki/Image:Japan_Kyoto_KiyoMizuDera_pagoda_DSC00616.jpg; creative commons license.

[48] L.A. Paquette, J.C. Stowell, *J. Am. Chem. Soc.* **93** (1971) 2459.
[49] http://commons.wikimedia.org/wiki/Image:Pig_USDA01c0116.jpg; public domain.
[50] C.J. Pederson, *J. Am. Chem. Soc.* **89** (1967) 7017.
[51] G.W. Gokel, D.M. Dishong, C.J. Diamond, *J. Chem. Soc., Chem. Commun.* (1980) 1053.
[52] J.R. Beadle, G.W. Gokel, *Tetrahedron Lett.* **25** (1984) 1681.
[53] S. Shinkai, M. Ishihara, K. Ueda, O. Manabe, *J. Chem. Soc., Chem. Commun.* (1984) 727.
[54] V.J. Gatto, G.W. Gokel, *J. Am. Chem. Soc.* **106** (1984) 8240.
[55] P.E. Eaton, B.D. Liepzig, *J. Am. Chem. Soc.* **105** (1983) 1656.
[56] J.A. Marshall, M. Lewellyn, *J. Am. Chem. Soc.* **99** (1977) 3508.
[57] M. Nakazaki, K. Yamamoto, M. Maeda, *Chem. Lett.* (1981) 1035.
[58] *Chemistry* (1967) July-Aug, p. 37.
[59] W.D. Ollis, C. Smith, D.E. Wright, *Tetrahedron*, **35** (1979) 105.
[60] D.E. Nettleton Jr., D.M. Balitz, T.W. Doyle, W.T. Bradner, D.L. Johnson, F.A. O'Herron, R.H. Schreiber, A.B. Coon, J.E. Moseley R.W. Myllymaki, *J. Nat. Prods.* **43** (1980) 242.
[61] J. Aronson, *Brit. Med. J.*, **319** (1999) 7215.
[62] T.W. Doyle, D.E. Nettleton, R.E. Grulich, D.M. Balitz, D.L. Johnson A.L. Vulcano, *J. Am. Chem. Soc.* **101** (1979) 7041.
[63] J. Guilhem, A.; Ducruix, C. Riche, C. Pascard, *Acta crystal.* **B32** (1976) 936.
[64] G. Mellows, P.G. Mantle, T.C. Feline, D.J. Williams, *Phytochem.* **12** (1973) 2717.
[65] J.L. Gaston, M.F. Grundon, *J. Chem. Soc. Perkin I*, (1980) 2294.
[66] H.-D. Martin, B. Mayer, M. Putter, H. Hochstetter, *Angew. Chem. Int. Ed. Engl.* **20** (1981) 677.
[67] Drawing of a Pterandon, artist Arthur Weasley, Wikimedia Commons, Creative Commons License, http://commons.wikimedia.org/wiki/Image:Pteranodon_BW.jpg
[68] J.G. Henkel, L.A. Surlock, *J. Am. Chem. Soc.* **95** (1973) 8339.
[69] S. Weinstein, *Experientia, Suppl. II* (1955) 137.
[70] A. Nickon, T. Iwadare, F.J. McGuire, J.R. Mahajan, S.A. Narang, B. Umezawa, *J. Am. Chem. Soc.* **92** (1970) 1688.
[71] D.S. Tarbell, A.T. Tarbell, *Roger Adams: Scientist and Statesman* (American Chemical Soc., Washington D.C., 1981), p63.
[72] M. Doyle, W. Parker, P.A. Gunn, J. Martin, D.D. MacNicol, *Tetrahedron Lett.* (1970) 3619.
[73] N.J. Leonard, J.C. Coll, *J. Am. Chem. Soc.* **92** (1970) 6685.
[74] W. Kliegel, D. Nanninga, U. Riebe, S.J. Rettig, J. Trotter, *Can. J. Chem.* **72**, (1994), 1735.
[75] J.Y. Li, G.A. Strobel, *Phytochem.* **57** (2001) 261.
[76] W.D.S. Motherwell, N.W. Isaacs, O. Kennard, I.R.C. Bick, J.B. Bremner, J. Gillard, *Chem. Commun.* (1971) 133.
[77] P.L. Katavic, M.S. Butler, R.J. Quinn, P.I. Forster, G.P. Guymer, *Phytochem.* **52** (1999) 529.
[78] C.G. Arena, F. Faraone, M. Fochi, M. Lanfranchi, C. Mealli, R. Seeber, A. Tiripicchio, *J. Chem. Soc., Dalton Trans.*, (1992) 1847.
[79] A.R. Carroll, J.C. Coll, D. J. Bourne, J. K. MacLeod, T. M. Zabriskie, C. M. Ireland, B. F. Bowden, *Aust. J. Chem.* **49** (1996) 659.
[80] Y. Asakawa, K. Takikawa, M. Toyota, T. Takemoto, *Phytochem.*, **21**, (1982) 2481.
[81] F. Orallo, M. Lamela, M. Camiña, E. Uriarte, J.M. Calleja, *Planta Med.* **64** (1998) 116.
[82] V.S. Kamat, F.V. Chuo, I. Kubo, K. Nakanishi, *Heterocycles* **15**, (1981) 1163.
[83] M. Hite, W. Rinehart, W. Braun, H. Peck, *Am. Ind. Hyg. Assoc. J.* **40** (1979) 600.
[84] J.S. Yadav, E.J. Reddy, T. Ramalingam, *New J. Chem.* **25** (2001) 223.
[85] J. Meinwald, L. Hendry, *Tetrahedron Lett.*, (1969) 1657.
[86] http://en.wikipedia.org/wiki/Carmustine
[87] http://www.nicnas.gov.au
[88] K. K. Chexal, J.P. Springer, J. Clardy, R.J. Cole, J.W. Kirksey, J.W. Dorner, H.C. Cutler, B.J. Strawter, *J. Am. Chem. Soc.* **98** (1976) 6728.
[89] P.G. Staksly, M.E. Schlosser, *J. Biol. Chem.* **161** (1945) 513.
[90] http://commons.wikimedia.org/wiki/Image:Cow_K5176-3.jpg, photo: Bian Schack, public domain.
[91] T.-y. An. L.-h. Hu, Z.-l. Chen, K.-Y. Sim, *Tetrahedron Lett.* **43** (2002) 163.
[92] T.A. Smitka, F. Bonjouklian, L. Doolin, N.D. Jones, J.B. Deeter, W.Y. Yoshida, M.R. Prinsep, R.E. Moore, G.M. L. Patterson, *J. Org. Chem.* **57** (1992) 857.
[93] E.I. Graziani, T.M. Allen, R.J. Andersen, *Tetrahedron Lett.*, **36** (1995) 1763.
[94] H. Speisky, B.K. Cassels, *Pharm. Res.* **29** (1994) 1.
[95] R.R. Izac, W. Fenical, J.M. Wright, *Tetrahedron Lett.* **25** (1984) 1325.

[96] W. E. Billups and Michael M. Haley, *J. Am. Chem. Soc.* **113** (1991) 5084.
[97] D.S. Matteson, W.C. Hiscox, L. Fabry-Asztalos, G.-Y. Kim, W.F. Siems, III, *Organometallics*, **20** (2001) 2920.
[98] http://en.wikipedia.org/wiki/Arachidonic_acid
[99] http://en.wikipedia.org/wiki/Warfarin
[100] V.M. Molotov, "*Molotov Remembers*" (1992, Russia), also see the article in the NY Times: http://query.nytimes.com/gst/fullpage.html?res=9E0DE6D7143FF936A35750C0A9659C8B63
[101] S. Marumo, M. Nukina, S. Kondo, K. Tomiyama, *Agric. Biol. Chem.*, **46** (1982) 2399.
[102] A. Al-Azmi, A.-Z.A. Elassar, B.L. Booth, *Tetrahedron* **59** (2003) 2749.
[103] J. Meinwald, Q. Huang, J. Vrkoč, K.B. Herath, Z.-C. Yang, F. Schröder, A.B. Attygalle, V.K. Iyengar, R.C. Morgan, T. Eisner, *Proc. Natl. Acad. Sci. USA*, **95** (1998) 2733.
[104] G. Jimenez-Bueno, G. Rapenne, *Tetrahedron. Lett.* **44** (2003) 6261.
[105] http://commons.wikimedia.org/wiki/Image:Scanning_Electron_Micrograph_of_a_Flea.jpg, public domain
[106] A. Shulgin, A. Shulgin, *PiHKAL: A Chemical Love Story* (Berkeley, California, Transform Press, 1991).
[107] M.R. Guasch-Jané, C. Andrés-Lacueva, O. Jáuregui, R.M. Lamuela-Raventós, *J. Archaeol. Sci.* **33**, (2006) 98.
[108] L.M. Stark, X.-F. Lin L.A. Flippin, *J. Org. Chem.*, **65** (2000) 3227.
[109] F. Xia, D. Nagrath, S.M. Cramer, *J. Chromatogr.* **A989** (2003) 47.
[110] S.H. Chanteau, J.M. Tour, *J. Org. Chem.*, **68** (2003) 8750.
[111] Y. Shirai, A.J. Osgood, Y. Zhao, K.F. Kelly, J.M. Tour, *Nano Lett.* **5** (2005) 2330.
[112] R. Misra, R.C. Pandey, S. Dev, *Tetrahedron* **35** (1979) 2301.
[113] http://en.wikipedia.org/wiki/Thomas_Hardwicke
[114] T. Yabuta, *J. Chem. Soc., Trans.*, **125** (1924) 575.
[115] http://en.wikipedia.org/wiki/Cyclic_adenosine_monophosphate
[116] M. Heinrich, J. Barnes, S. Gibbons, E. Williams, *Fundamentals of Pharmacognosy and Phytotherapy* (Churchill Livingstone, 2004, pg 165).
[117] P.K. Pokhrel, K.V. Ergil, *Clinical Acupuncture and Oriental Medicine* **1** (2000) 161.
[118] http://commons.wikimedia.org/wiki/Image:Spiro_Agnew.jpg, public domain.
[119] John L. Meisenheimer, Sr., Professor of Chemistry Emeritus, Eastern Kentucky University, personal communication.
[120] D.H. Nouri, D.J. Tantillo, *Curr. Org. Chem.* **10** (2006) 2055.
[121] J. Lagona, P. Muchophadyay, S. Chakrabarti, L. Isaacs, *Angew. Chem. Int. Ed.* **44** (2005) 4844.
[122] Y. Saikawa, K. Moriya, K. Hashimoto, M. Nakata, *Tetrahedron Letters* **47** (2006) 2535.
[123] M. Musman, I.I. Ohtani, D. Nagaoka, J. Tanaka, T. Higa, *J. Nat. Prod.* **64** (2001) 350.
[124] R.D. Hunt, J.T. Yustein, L. Andrews, *J. Chem. Phys.*, **98** (1993) 6070.
[125] http://commons.wikimedia.org/wiki/Image:Tempeh_tempe.jpg, public domain.
[126] P.J.F. Henderson, H.A. Lardy, *J. Biol. Chem.* **245** (1970), 1319.
[127] S. Urban, J.W. Blunt, M.H.G. Munro, *J. Nat. Prod.*, **65** (2002) 1371.
[128] D.W. Ball, *J. Mol. Struc. (Theochem)* **626** (2003) 217.
[129] A. Ault, *J. Chem. Educ.* **78** (2001) 924.
[130] H.E. Zimmerman, G.L. Grunewald, R.M. Paufler, M.A. Sherwin, *J. Am. Chem. Soc.* **91** (1969) 2330.
[131] http://en.wikipedia.org/wiki/Image:Chav.jpg, public domain image, drawn by J.J. McCullough.
[132] See http://www.avantilipids.com for info on both *DOPE* and *DOGs*.
[133] S. Nozoe, J. Furukawa, U. Sankawa, S. Shibata, *Tetrahedron Lett.* **17** (1976) 195.
[134] H. Shimogawa, Y. Kwon, Q. Mao, Y. Kawazoe, Y. Choi, S. Asada, H. Kigoshi, M.A. Uesugi, *J. Am. Chem. Soc.* **126** (2004) 3461.
[135] http://en.wikipedia.org/wiki/Forskolin
[136] S.H. Toma, M. Uemi, S. Nikolaou, D.M. Tomazela, M.N. Eberlin, H.E. Toma, *Inorg. Chem.* **43** (2004) 3521.
[137] K. Huvaere, M.L. Andersen, M. Storme, J. Van Bocxlaer, L.H. Skibsted, D. De Keukeleire, *Photochem. Photobiol. Sci.*, 5 (2006) 961.
[138] A. Rother, A.E. Schwart, *J. Chem. Soc. D.* (1969) 1411.
[139] J.P. Ferris, R.C. Briner, C.B. Boyce, *J. Am. Chem. Soc.* **93** (1971) 2958.
[140] E. Spath, F. Kuffner, *Ber. Deutch Chem. Gess.* **67** (1934) 55.
[141] M. Yoshida, M. Ezaki, M. Hashimoto, M. Yamashita, N. Shigematsu, M. Okuhara, M. Kohsaka, K. Horikoshi, *J. Antibiotics* **43** (1990) 748.
[142] A.G.M. Barrett, K. Kasdorf, G.J. Tustin, D.J. Williams, *J. Chem. Soc., Chem. Commun.* (1995) 1143.
[143] J. Baker, A. Kessi, B. Delley, *J. Chem. Phys.* **105** (1996) 192.
[144] http://www.chm.bris.ac.uk/motm/htx/htx_h.htm

[145] M. Murata, M. Kumagai, J.S. Lee, T. Yasumoto, *Tetrahedron Lett.* **28** (1987) 5869.
[146] http://en.wikipedia.org/wiki/Oseltamivir
[147] D.B. Amabilino, P.R. Ashton, A.S. Reder, N. Spencer, J.F. Stoddart. *Angew. Chem. Int. Ed. Engl.* **33** (1994) 1286.
[148] http://webbook.nist.gov/cgi/cbook.cgi?Name=fart*&Units=SI&cTG=on&cTC=on&cTP=on&cTR=on
[149] F.A. Cotton, J.P. Donahue, C.A. Murillo, *J. Am. Chem. Soc.* **125** (2003) 5436.
[150] S.P. Gunasekera, M. Gunasekera, R.E. Longley, *J. Org. Chem.* **55** (1990) 4912.
[151] Image is Elizabeth I of England, the Armada Portrait, Woburn Abbey (George Gower, *ca* 1588), public domain: http://commons.wikimedia.org/wiki/Image:Elizabeth_I_%28Armada_Portrait%29.jpg.
[152] http://commons.wikimedia.org/wiki/Image:NCI_Visuals_Food_Hamburger.jpg, public domain, photographer Len Rizzi:
[153] A. Yokoyama, T. Kojima, K. Ohkubo, S. Fukuzumi, *Chem. Commun.*, (2007) 3997.
[154] D. Yang, G. Meng, S. Zhang, Y. Hao, X. An, Q. Wei, M. Yea, L. Zhang, *Chem. Commun.*, (2007) 1733.
[155] S. Trofimenko, *Scorpionates: Polypyrazolylborate Ligands and Their Coordination* Chemistry, (World Scientific, Singapore, 1999).
[156] http://www.chm.bris.ac.uk/motm/sscorpionine/sscorpionine.htm
[157] http://commons.wikimedia.org/wiki/Image:Penicilliummandarijntjes.jpg, GNU Free document license.

2. Minerals with Silly or Unusual Names

There are hundreds of minerals, and a surprisingly large number of them have silly names. Most minerals are named after either the locality where they were first found or the person who discovered them. I've only collected together a few of my favourites here, but comprehensive lists of the others can be found on the web [1,2,3].

Cummingtonite

This mineral must have the silliest name of them all ! It is a *silicate* rock containing magnesium and iron, with the basic formula $(Mg,Fe)_7Si_8O_{22}(OH)_2$. It got its name from the locality where it was first found, Cummington, Massachusetts, USA.

Photo of *cummingtonite.*
(Copyright Getty Images, DEA picture library, used with permission).

Arsenolite

Arsenolite from St Etienne, France.

(Photo copyright: Robert Meyer [4], used with permission).

This is a naturally occurring mineral, whose correct name is *cubic arsenic trioxide* (As_2O_3). It is also the primary product whenever arsenic ores are smelted, and is used in industry as a glass decolourising agent. Another related mineral with a similar silly name is *arsenolamprite*, which is a native form of arsenic.

Dickite

Photo of a large *dickite*, scale bar is 1 cm.

(Used with permission of Jeffrey G. Weissman [5]).

Dickite, $Al_2Si_2O_5(OH)_4$, is a (kaolin) clay-like mineral which exhibits mica-like layers with *silicate* sheets of 6-membered rings bonded to *aluminium oxide/hydroxide* layers. *Dickite* is used in ceramics, as paint filler, rubber, plastics and glossy paper. It got its name from the geologist that discovered it around the 1890s, Dr. W. Thomas Dick, of Lanarkshire, Scotland.

Fukalite

Photo of *fukalite.*
(Used with permission of Gerhard Brandstetter[6]).

This wonderfully named mineral gets its name from the Fuka mine in the Fuka region of southern Japan, where it is mined by the local Fukas. It is very rare, and is a form of *calcium silico-carbonate*, with formula $Ca_4Si_2O_6(CO_3)(OH,F)_2$.

Kinoshitalite

(Image copyright Shinichi Kato,[7] used with permission).

Although it sounds like the trade name of a laxative, this is a type of mica with formula $(Ba,K)(Mg,Mn)_3Si_2Al_2O_{10}(OH)_2$ and is found in Japan and Sweden.

It is green and vitreous, and apparently is about as hard as fingernails. Its name translates as the Japanese words for "under the tree" (*ki* = tree; *no* = possessive particle; *shita* = under), but it's actually named after Dr Kameki Kinshita who was a Japanese mineralogist.

Kunzite (Spodumene)

A bunch of kunzite crystals. (Copyright Dave Bathelmy, origin webminerals.com, and used with permission).

This mineral is a pink (of course) gemstone, named after the gemologist G.F. Kunz who described it in 1902. *Kunzite* is a fragile stone, which shows different shades of colour when viewed from different directions. Called an "evening" stone, it should not be exposed to direct sunlight which can fade its color in time.

Its alternative name, *spodumene*, sounds like an American shop that sells computer nerds ("Spod-U-Mean"). This comes from the Greek *spodoumenos* meaning 'burnt to ashes', since in this form it has an ashen colour.

Fornacite

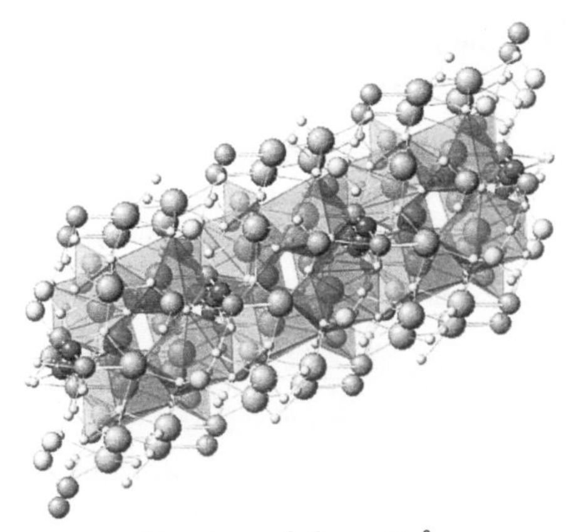

Structure of *fornacite* [8].

This is a mineral that is composed of a basic *chromate-arsenate* compound of Pb and Cu with formula: $(Pb,Cu^{2+})_3[(Cr,As)O_4]_2(OH)$. If it could be polished into a gemstone, it sounds ideal for a ring that a cheating husband might buy his mistress.

Aminoffite

Ah'm in a fight...
Ah'm in a fight, too. [9].

A boxer's favourite mineral? This mineral has formula $Ca_3Be_2(Si_3O_{10})(OH)_2$ and forms clear pyramidal crystals. It's named after the Swedish Mineralogist Gregori Aminoff.

Carbuncle

A *pyrope* gem, certainly not a carbuncle! [10]

This mineral has got nothing to do with skin lesions or sores, except for the red colour. It's an old term that used to refer to any fiery red mineral, but nowadays is restricted to red garnets such as *pyrope*.

Aenigmatite

This mineral gets its 'enigmatic' name from the fact that its chemical composition was originally difficult to determine. It is an iron and titanium silicate with sodium as a charge-balancing cation[11].

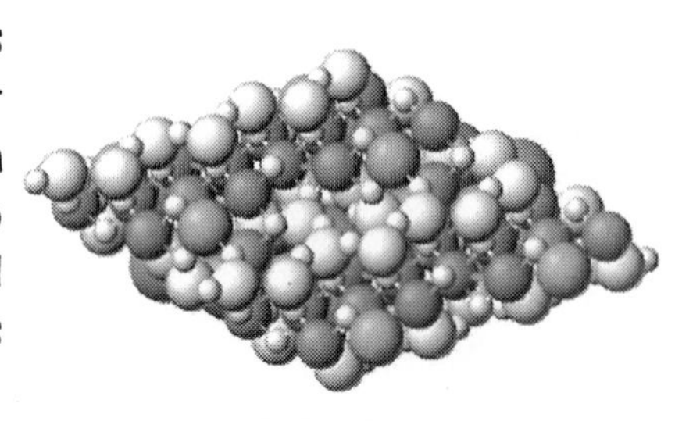

The structure of *aenigmatite* [12] - or is it?

Because it does not easily fit into the current classification system, it is classified as an 'Unclassified Silicate' - if that makes any sense.

Apatite

A large apatite.
(Copyright Getty Images,
Photo: Visuals Unlimited, used with permission).

A mineral for hungry people? *Apatite* is a phosphate mineral with the composition $Ca_5[PO_4]_3(OH,F,Cl)$. It has been used extensively as a phosphorus fertilizer and is still mined for that purpose today. The mineral called *asparagus stone* is a appropriately a type of green *apatite*.

It's called *apatite* from the Greek word *apate*, meaning 'deceit', referring to the fact that this mineral is easily mistaken for others such as *beryl* and *tourmaline*. Ironically, *apatite* is the mineral that makes up the teeth in all vertebrate animals as well as their bones.

Carnallite

Carnallite is $KMgCl_3\ 6H_2O$, an evaporite mineral. Surprisingly, for a mineral called *carnallite*, it doesn't exhibit any cleavage... It's used as an ore for potassium fertilizers, and is named after Rudolf von Carnall, a Prussian mining engineer, whose (carnall?) knowledge of the subject was famous.

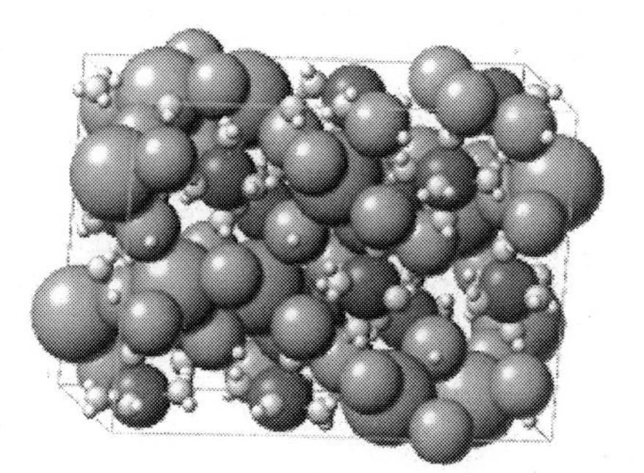

Structure of *carnallite*[13]).

Welshite

This wonderfully named mineral is called after the US amateur mineralogist (and science teacher, and auxiliary police officer, and motorcycle enthusiast) Wilfred R. Welsh. Its chemical formula is $Ca_2SbMg_4FeBe_2Si_4O_{20}$. Some people think it's quite a nice mineral, but others think it's 'well-shite'.

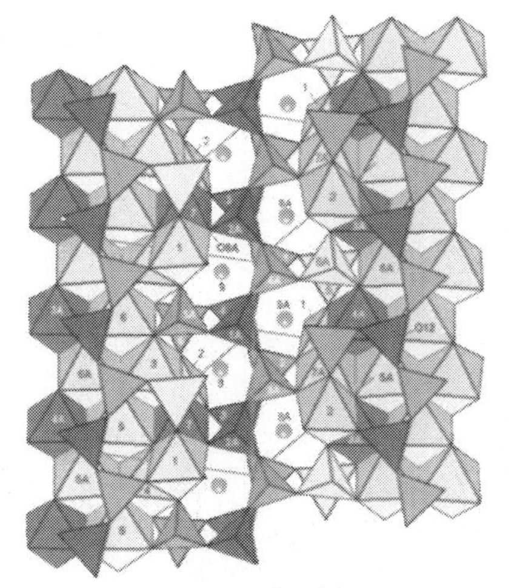

Structure of welshite.
(used with permission from *American Mineralogist*[14])

With the way minerals are named, it's lucky that none have been discovered by geologists named Pylosh, Baggosh, Crokkosh, or even Lumposh!

Parisite

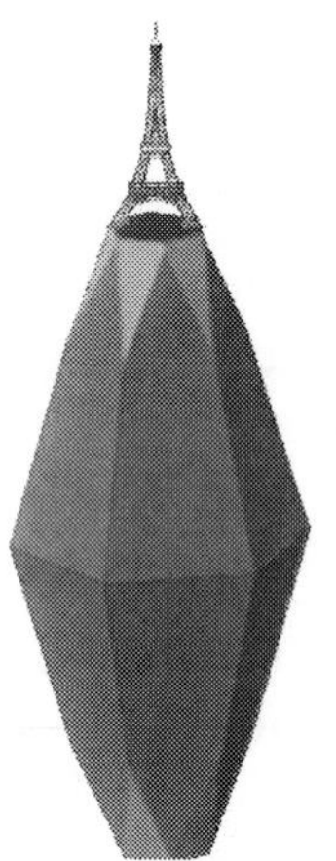
If it's *Paris-ite*, there should be an Eiffel Tower [15].

This mineral sounds like it grows on other minerals. Or is it what the Eiffel Tower is made from? Actually, it has nothing to do with the city Paris (nor even Paris Hilton), but it's named after J.J. Paris, who was a mine proprietor at Muzo, north of Bogota, Columbia. It is a member of the *Bastnasite* group of complex *carbonate-fluoride* minerals, with the chemical formula $Ca(Nd,Ce,La)_2(CO_3)_3F_2$. The crystals usually form as acute double hexagonal pyramids with flattened ends.

Clintonite

Bill Clinton [16]

This mineral sounds like it should be found in close proximity to *internite* or *lewinskyite*... but it was actually named after American statesman De Witt Clinton (1769-1828), not the notorious ex-President (or his wife). It's a *calcium magnesium aluminosilicate hydroxide* mineral, and it's related to *margarite*. It is found abundantly in the northern extent of New York - ironically the State of which Hilary Clinton is a junior senator.

Crocidolite

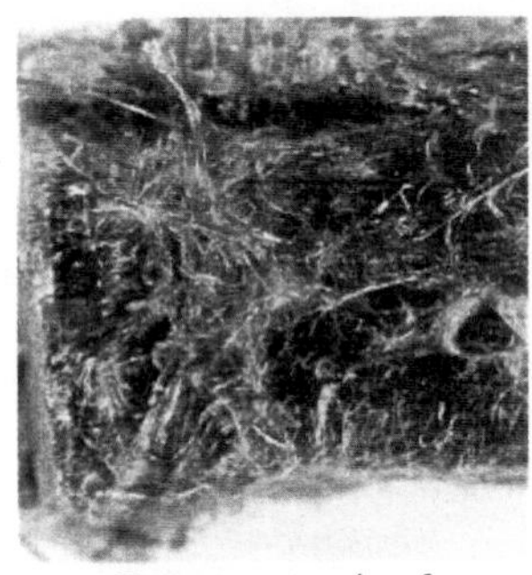

I want a sample of *crocidolite*... and make it snappy!
(Photo: John Hayman [17]).

This has nothing to do with crocodiles, but is actually the name of the mineral form of blue asbestos, with chemical formula $Na_2Fe_5Si_8O_{22}(OH)_2$, and is the most lethal form of all the asbestos types. It was first described in 1815 by M.H. Klaproth under the name *Blaueisenstein* (blue ironstone), and later in 1831 by J.F. Hausmann, who gave it its present name based on the Greek words for 'woolly' (*krokis*) and rock (*lithos*) on account of its fibrous-like appearance [18].

Oh no, I just *shattuckite*!
(Photo: Dave Dyet [19])

Shattuckite

This mineral sounds painful, but is actually just named after the locality in which it is found, the Shattuck mine in Arizona. Its actual formula is $Cu_5Si_4O_{12}(OH)_2$.

Khanneshite

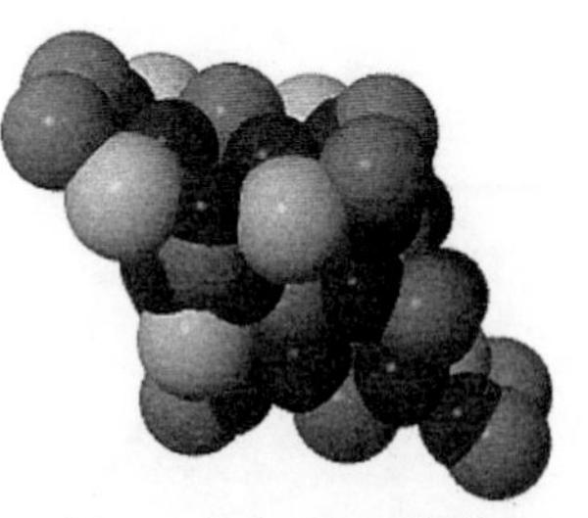
A box of atoms, or is it a *khanneshite*? [20]

Maybe this mineral is made from *constipatic acid*! But it's amazing how this name got past the International Mineralogical Association when they approved the new mineral name. It was named after the discovery locality at Khanneshin, Afghanistan. All minerals have the suffix '-ite'. and so the name they gave to this mineral is - *khanneshite*. I can only presume that there were no Scots on the committee that approved the name... In German it's even better, as the last 'e' is removed to get *khanneshit*. Well, I suppose if you can get a *vaginatin*, you can also get a *khanneshit*.

Microlite

Microlite is not one of the components of a small airplane, but is a *tantalum/niobium oxide* mineral that can be slightly radioactive. Its correct chemical formula is: $(Ca, Na)_2Ta_2O_6(O, OH, F)$, and its name means 'small rock' in Greek [21].

Microlite (the plane, not the mineral)
(Copyright Getty Images, used with permission, photographer: Dorling Kindersley).

Stichtite

A sticky mineral?
(Photo: Dave Dyet[22]).

Stichtite is a lilac coloured mineral which is a *hydrated magnesium chromium carbonate hydroxide*. This is seen fairly commonly as streaks and small lenses in the green *serpentine* in the metamorphic rocks of Western Tasmania, but is very rare elsewhere. It was named after Robert Sticht, a director of a mining company.

Coalingite

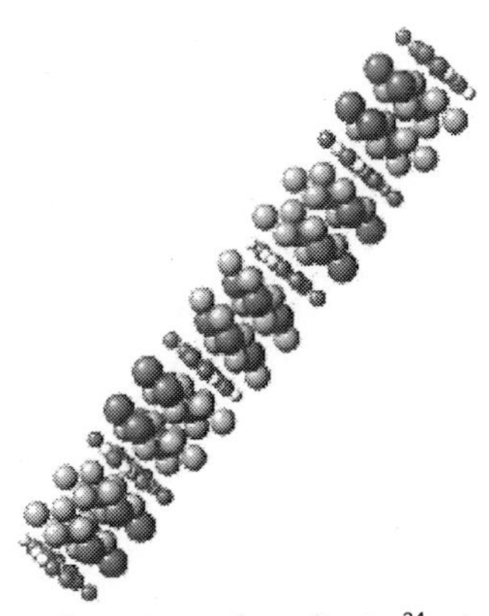

Structure of *coalingite*[24].

This superbly named mineral takes some licking... It is actually an interstratified *hydroxyl carbonate*, but whenever anyone says its name, they normally lower their voices for the rest of the discussion, probably because it sounds like a contraction of *coitus* and *lingus*. But it was named after being found in the vicinity of the town of Coalinga, California, which was itself named after 'Coaling Station A'[23].

Snottites

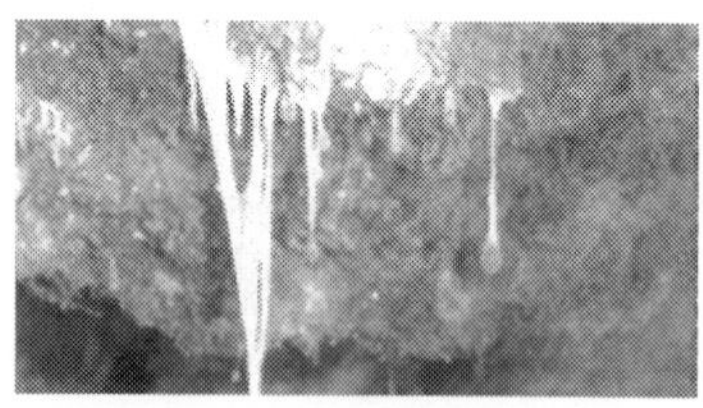
Is this what happens when caves catch a cold?
(Photo: Kenneth Ingham[26]).

These are gelatinous, dripping microbial draperies ('mucus stalactites') composed of elemental sulfur, iron oxide crusts, gypsum, and densely packed bacteria, and are found in caves [25]. They are part microbe, part mineral and are formally known as *biovermiculations*, although *snottite* is more descriptive.

Snottites are produced by sulfurphilic micro-organisms and drip *sulfuric acid* with a pH of 0.3 to 0.7. Other microbial structures include "blue goo", which are lavender structures attached to the walls of the cave, and "red goo," a complex clay breakdown product containing clusters of bacterial cells and having a pH ranging from 3.9 to 2.5. Other microbial stalactites go by the fancifully names of "phlegm balls," "green slime," "punk rocks," "hairy sausages," "slime balls," and "beads on a string."

On a similar theme we also have: *coprolites*, which are fossil feces, *regurgitalites*, which are fossil vomit or pellets, and *cololites*, fossilised stomach, gut, or colon contents.

Burpalite

This mineral with the wonderful name of *burpalite*, $Na_2CaZrSi_2O_7F_2$, is named after the Burpala massif in Buryatia, Russia [27]. It sounds a bit like a medicine you give to babies to relieve wind.

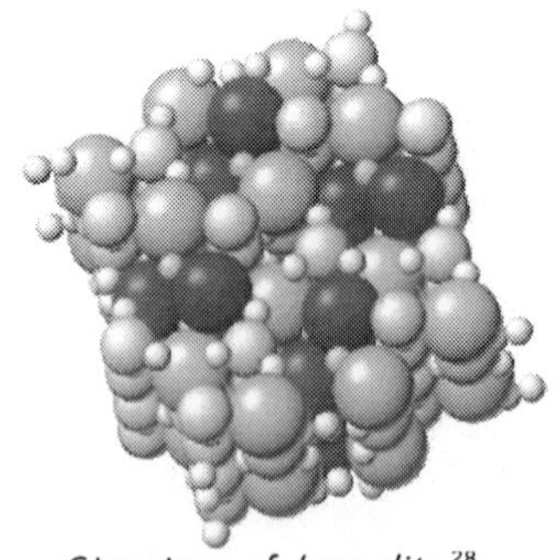
Structure of *burpalite* [28].

Sillimanite

A large piece of sillimanite.
(Photo: USGS[29]).

This mineral wasn't named after the clumsy fool that tripped over it, but was named in honour of the American mineralogist Benjamin Silliman, who was a chemistry professor at Yale. It is a form of *aluminium silicate*, with no real value, except in Idaho, where the Clearwater River Valley has *sillimanite* cobbles that are carved into figurines and sold as souvenirs of Idaho.

Kutnahorite

This is another mineral that geology students love to mispronounce as "cuttin' a whore right". It's a $CaMn(CO_3)_2$ mineral originally from Czechoslovakia, and described as a massive and granular material occurring in veins. Its color is almost always some shade of pink, with well-developed cleavage with cleavage surfaces that are commonly curved (aren't they always?).

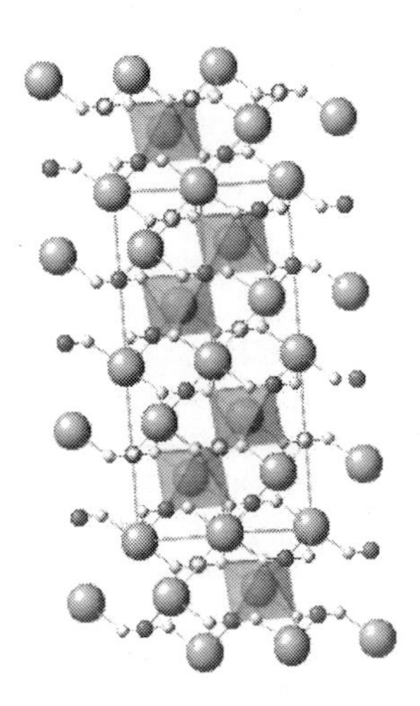

Right: Structure of kutnahorite[30].

Soddyite

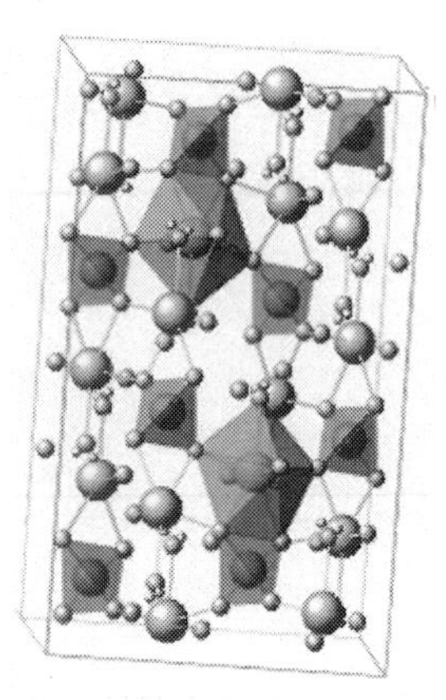

This is a *silicate-uranyl* (glass toilet?) mineral with formula $(UO_2)2SiO_4 \cdot 2H_2O$ that was named after Frederick Soddy, the British physicist and radiochemist - who also coined the term 'isotope'. In German it's known as *soddyit* which is probably what the geologist would exclaim when they stubs their toe on this rock!

Left: Soddyite structure [31].

Jimthompsonite

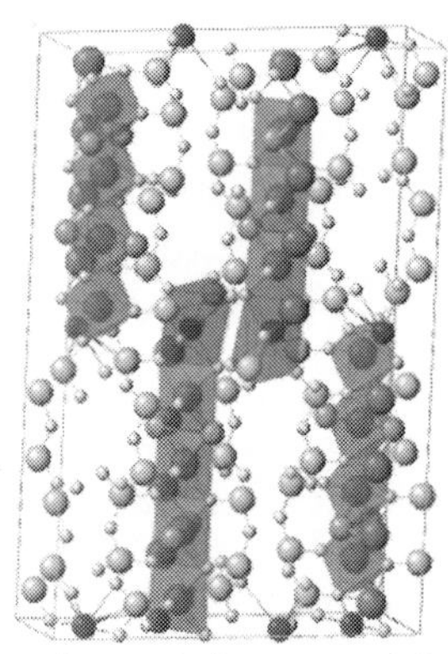

The crystal structure of *jimthompsonite* [32].

This mineral is a mixture of *iron* and *magnesium silicates* - with formula $(Mg;Fe^{2+})_5Si_6O_{16}(OH)_2$ - that's found in a talc quarry, near Chester, Vermont, USA. It was named after James Burleigh Thompson, Jr. [33], who was an eminent petrologist at Harvard University in the 1940s and 1950s. This is a different Jim Thompson to the one who helped establish the silk industry in Thailand [34]. There's a monoclinic version of this mineral called *clinojimthompsonite*.

Labradorite

A piece of *labradorite* showing labradorescence...?
(Photo: Herwig Kavallar [35])

A labrador dog exhibiting labradorescence...?
(Adapted from a photo by Herwig Kavallar [36])

Labradorite is a *silicate* mineral that is named after Labrador in eastern Canada, where it was first discovered. *Labradorite* can produce a colourful play of light across cleavage planes and in sliced sections called *labradorescence*, which sounds like a perfume based on the smell of old dogs, or maybe the glow from a radioactive dog!

Coffinite

This mineral has a very appropriate name, considering it's a *silicate* uranium ore that's highly radioactive. It's named after the American geologist Reuben Clare Coffin. I wonder if this mineral would react well with the *sarcophagene* or *sepulchrate* molecules mentioned earlier?

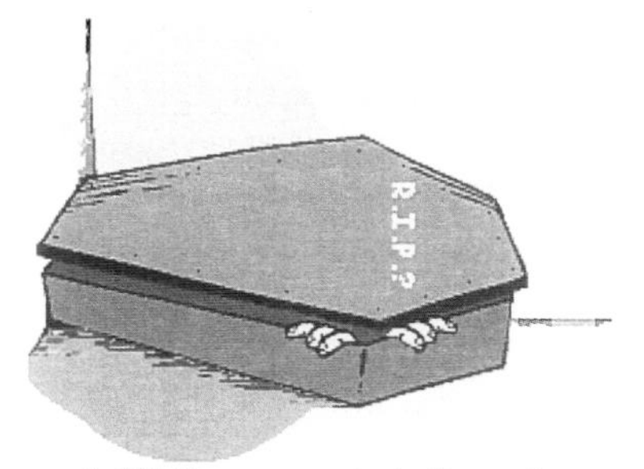

Coffinite - a vampire's favourite mineral?
(Image modified from one obtained from clipartheaven.com, used with permission)

Analcite

I thought it was time to show you my *analcite*! Although *analcite* is a valid name for this mineral, it normally goes by the less amusing name of *analcime*. It's a form of *sodium aluminium silicate*, and it gets its name from the Greek for 'weak', referring to a weak electrical charge developed on rubbing. So if you rub your *analcite*, you may get a shock...

Look at my nice *analcite*![37]

This rock's a fuchsite better than most other minerals...
(Photo: Ra'ike[38])

Fuchsite

Fuchsite is a mineral, and is the green form of *muscovite*, $KAl_2(AlSi_3O_{10})(F, OH)_2$. It is used as an ornamental stone, and apparently has perfect cleavage...

Noselite

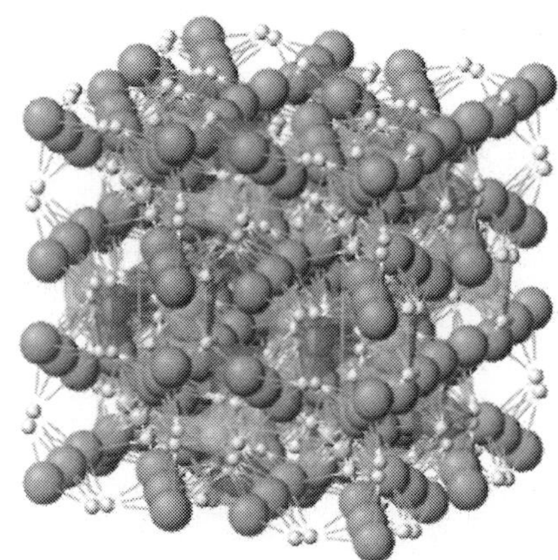

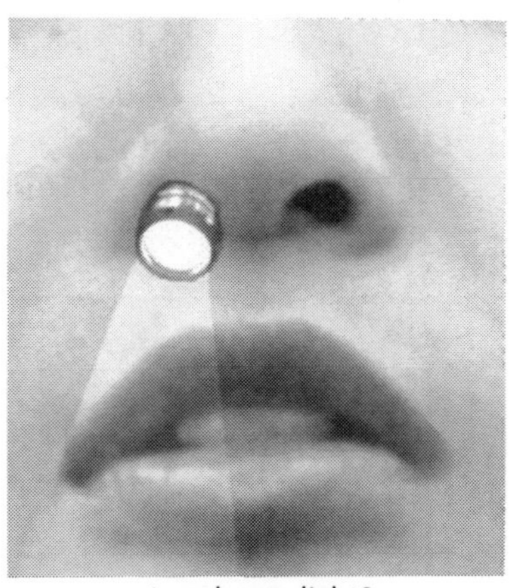

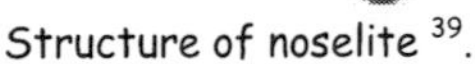

Structure of noselite [39].

A real nose-light?
(Photo: PWM)

A superb mineral name that sounds like just the sort of thing a geologist needs to find his way around in a dark cave. It's a *silicate* mineral with formula $Na_8[SO_4|Al_6Si_6O_{24}]$, and named after the German mineralogist Karl Wilhelm Nose, and also goes by the name of *nosean* - as in "I've just been *nosean* around this rock face". For experiments involving real nose-lights, where the researcher stuck lightbulbs up the noses of volunteers to see how it affected their visual performance, see the wonderfully bonkers paper in *Nature* by Wetherick [40].

Carlsbergite

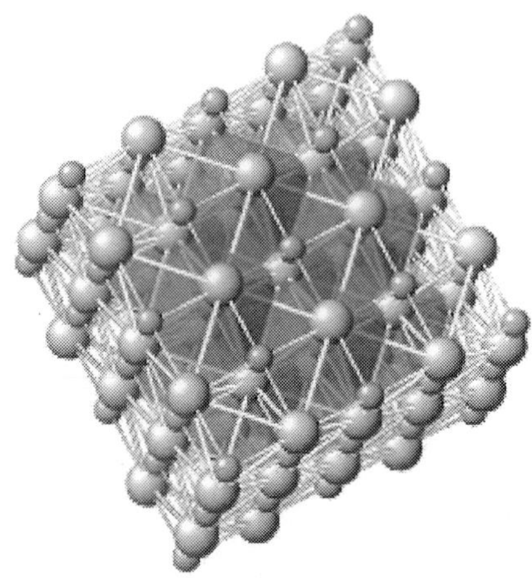

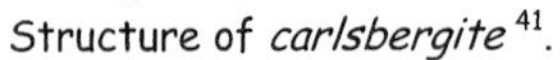
Structure of *carlsbergite*[41].

Oh, I thought you said a Carlsberg Lite?
(Photo: Hawyih [42])

This mineral was named after the Carlsberg Foundation of Copenhagen, Denmark (yes, the same one that makes the beer), which supported the recovery and cutting of the *Agpalilik* meteorite wherein it was discovered. On the same theme, there's a nickel-sulfide mineral called *millerite*, which isn't to be confused with *Miller-lite*, the beer.

Pigeonite

A rock pigeon
(Photo: Christian Jansky [43]).

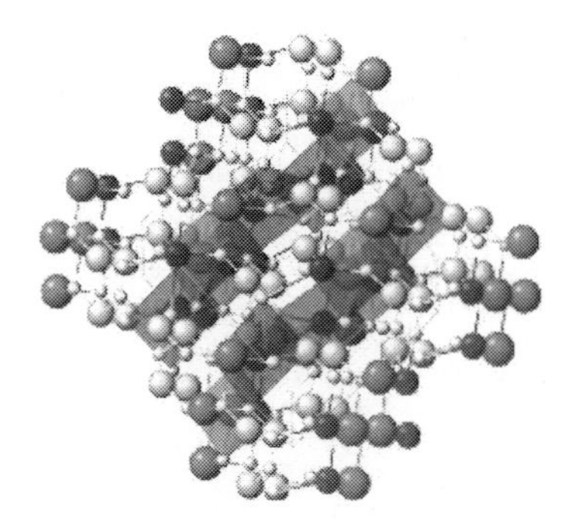

A pigeon rock! [44]

Pigeonite is a silicate mineral named after Pigeon Point in Minnesota where it was first found. It forms dark green crystals that are found in volcanic rocks on Earth and in meteorites from Mars or the Moon. Those darn pigeons get everywhere...

References

[1] http://www.webmineral.com
[2] http://www.mindat.org
[3] http://en.wikipedia.org/wiki/List_of_minerals_%28complete%29
[4] Robert O. Meyer, photo details http://www.mindat.org/photo-3113.html
[5] Jeffrey G. Weissman, *Photographic Guide to Mineral Species*, ed 2, (Excalibur Mineral Corp., Peekskill, NY, USA), and also at http://webmineral.com/specimens/picshow.php?id=1732
[6] http://austrianminerals.heim.at
[7] Shinichi Kato, '*Kato's Collections*': http://www.asahi-net.or.jp/~ug7s-ktu/english.htm
[8] Image created from data and the applet at: http://www.webmineral.com/data/Fornacite.shtml
[9] http://commons.wikimedia.org/wiki/Image:Ouch-boxing-footwork.jpg, photo: Cpl. M.L. Stiner, public domain.
[10] http://commons.wikimedia.org/wiki/Image:Piropas.jpg, photo: WesternDevil, GNU free document license.
[11] E. Cannillo, F. Mazzi, J.H. Fang, P.D. Robinson, Y. Ohya. *Am. Mineralogist* **56** (1971) 427.
[12] Image created from from data and the applet at: http://www.webmineral.com/data/Aenigmatite.shtml
[13] Image created from data and the applet at: http://www.webmineral.com/data/Carnallite.shtml
[14] E.S Grew, J. Barbier, J. Britten, U. Hålenius, C.K. Shearer. *American Minerologist*, **92**, (2007) 80.
[15] Image made from crystal structure data at: http://www.webmineral.com/data/Parisite-(Ce).shtml
[16] http://commons.wikimedia.org/wiki/Image:44_Bill_Clinton_3x4.jpg, public domain.
[17] http://commons.wikimedia.org/wiki/Image:Blue_asbestos_%28teased%29.jpg, public domain.
[18] R.P. Abratt, D.A. Vorobiof, N. White, *Lung Cancer* **45** (1) (2004) S3.
[19] http://commons.wikimedia.org/wiki/Image:Shattuckite_Hydrous_Copper_Silicate_Ajo_Pima_County_Arizona_1584.jpg; public domain.
[20] Image created from structure file held on the ISCD database at: http://icsdweb.fiz-karlsruhe.de/details.php?id%5B%5D=73133&PHPSESSID=734118bb8; based on data in: Yu.V. Belovitskaya, I.V. Pekov, E.R. Gobechiva, Yu.K. Schneider, *Kristallografiya* **47** (2002) 46.
[21] http://webmineral.com/data/Microlite.shtml
[22] http://commons.wikimedia.org/wiki/Image:Stichtite_on_serpentine_Basic_hydrous_magnesium_chromate_and_carbonate_New_Amianthus_Mine_Transvaal_South_Africa_1623.jpg; public domain.
[23] F.A. Mumpton, H.W. Jaffe, C.S. Thompson, *American Minerologist* **50** (1965) 1893.
[24] Image created from data and the applet at: http://www.webmineral.com/data/Coalingite.shtml
[25] http://en.wikipedia.org/wiki/Snottite
[26] http://commons.wikimedia.org/wiki/Image:Snottite.jpg; public domain.
[27] S. Merlino, N. Perchiazzi, A.P. Khomyakov, D.Y. Pushcharovsky, I.M. Kulikova, V.I. Kuzmin, *Eur. J. of Mineral.* **2** (1990) 177.
[28] Image created from data and the applet at: http://www.webmineral.com/data/Burpalite.shtml
[29] http://commons.wikimedia.org/wiki/Image:SillimaniteUSGS.jpg; public domain.
[30] Created from data in the applet at http://www.webmineral.com/data/Kutnohorite.shtml
[31] Created from data in the applet at http://webmineral.com/data/Soddyite.shtml
[32] Image created from data in the applet at: http://webmineral.com/data/Jimthompsonite.shtml
[33] E-N. Zen, *Am. Mineralogist*, **64** (1979) 6634.
[34] http://www.jimthompsonhouse.com
[35] http://commons.wikimedia.org/wiki/Image:Labradoryt%2C_Madagaskar.JPG, photo reproduced according to the GNU free documentation license.
[36] http://commons.wikimedia.org/wiki/Image:Afra_007.jpg; public domain.
[37] Photo reproduced from Wikimedia Commons under the GNU free document license, photographer: Linnell. http://commons.wikimedia.org/wiki/Image:Analcite.jpg
[38] http://commons.wikimedia.org/wiki/Image:Fuchsite.jpg; creative common licence: http://creativecommons.org/licenses/by-sa/2.5/
[39] Image made using data from the applet on webmineral.com; http://www.webmineral.com/data/Nosean.shtml
[40] N.E. Wetherick, *Nature* **266** (1977) 442.
[41] Image made using data from the applet on webmineral.com; http://www.webmineral.com/data/Carlsbergite.shtml
[42] http://commons.wikimedia.org/wiki/Image:Carlsberg_beer.jpg; public domain image.
[43] http://commons.wikimedia.org/wiki/Image:Columba_livia_02.jpg; Creative Commons Attribute ShareAlike License.
[44] Image made using data from the applet on webmineral.com; http://www.webmineral.com/data/Pigeonite.shtml

3. Proteins and Enzymes with Silly or Unusual Names

Luciferase and Luciferin

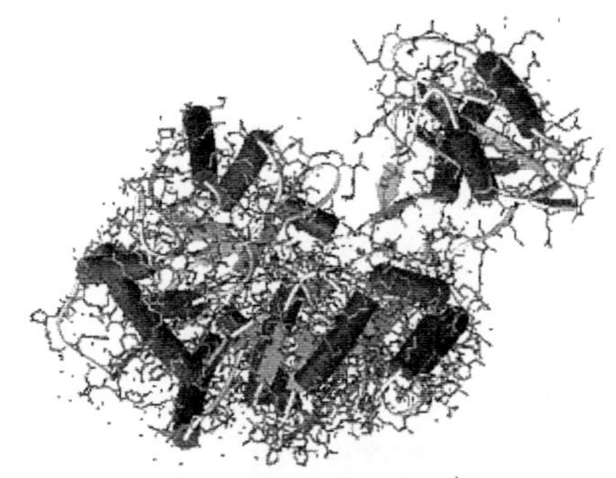

Luciferase structure[1]

Luciferase is an enzyme which reacts with *ATP* to cleave *luciferin*, its substrate. This cleavage reaction creates light, and is what causes the fiery glow in fireflies and certain types of fish. *Lucifer* actually means 'light bearing' in Latin.

Buccalin

This sounds like the molecule from which car seat-belts are made, but it's actually a neuropeptide which acts in nerves to stop *acetylcholine* release.

Gly-Met-Asp-Ser-Len-Ala
|
H_2N-Leu-Gly-Gly-Ser-Phe

Draculin

Christopher Lee as *Dracula* (or should that be *zinc finger protein Dracula*?).
(Copyright Getty Images, Silver screen collection, taken from the Hammer film: Dracula AD 1972, used with permission).

Draculin is the anticoagulant factor in vampire bat saliva [2]. Its full name is *zinc finger protein draculin* and it's a large *glycoprotein* made from a sequence of 411 *amino acids* [3].

Golf proteins

These are a class of proteins called *G-type proteins*, some of which are linked to the *olf*actory system - hence the name *Golf proteins* [4]. They help trigger the biochemical synthesis of neurotransmitters, which eventually leads to signalling, and gives us a sense of smell. They are known to have been present in the flatworms of the Precambrian era over 800 million years ago, which makes them one of the earliest known proteins.

Anyone for *Golf*? [5]

Profilactin

Actin is an intracellular fibre protein (best known in muscle contraction), and *profilin* is a protein that interacts with *actin* in order to "promote filament" formation. When *profilin* and *actin* are bound together, the complex was originally labelled *profilactin*, which is most appropriate since it was first isolated in sea urchin sperm [6].

Prophylactics - to stop the spread of *profilactin*?
(Copyright Getty Images, used with permission, photographer Ian Logan)

However, I'm told this name was not officially recognised. Also, the first submission of a name for this protein was *screw-in*... because when the filament is ejected from the tip of the sperm, the globular *actin* shoots outward in a screw-like motion. Although this, too, was a clever name given the protein's location, it was turned down as well. Its official name is now 'the *profilin-actin complex*'.

Beta-BuTX

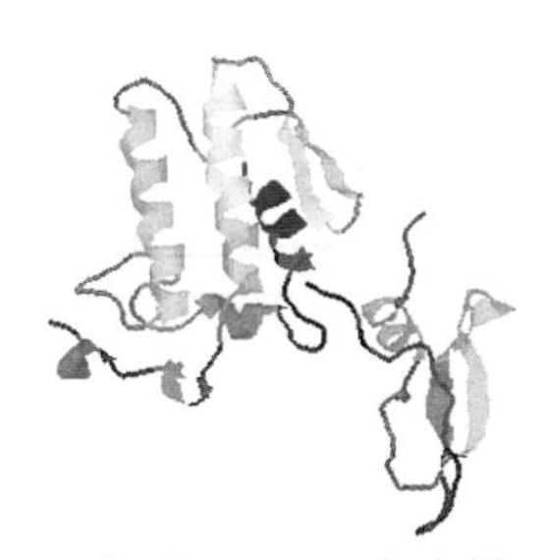

Your ButX, too, can look like these! [7].

This molecule sounds like the sales pitch for an exercise regime (get *better buttocks* using this!), but it's actually a snake venom with full name *beta-Bungarotoxin*) [8]. Maybe the venom is more potent if the snake bites you on the *ButX*.

ASS

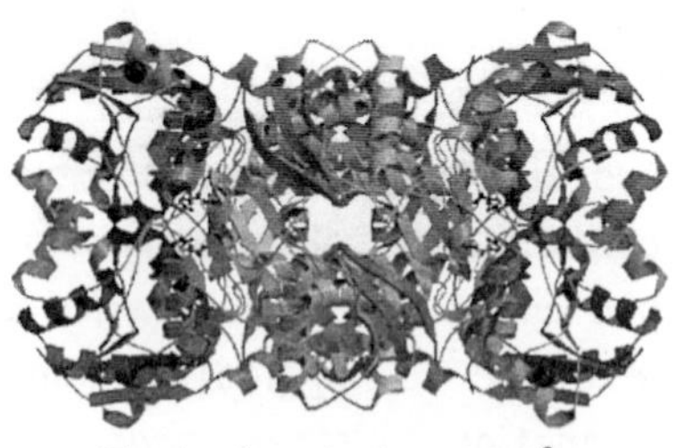
The back end of an *ASS*?[9]

This is the commonly used acronym for *argininosuccinate synthetase*, which is a chemical found in the brain. So it seems some people really do think with their *ASSes*.

Horseradish Peroxidase

Anyone for enzyme sauce?
(Photograph: PWM).

This is the version of the *peroxidase* enzyme that is isolated from the horseradish plant. Like all *peroxidases*, it converts harmful *peroxide* molecules (H_2O_2) into water molecules (H_2O). It is also used to label proteins and nucleic acids as an alternative to radio labelling. The molecule to be labelled is attached to the *horseradish peroxidase* molecule, and the mixture is then exposed to a substrate that changes from clear to coloured when it is oxidized by *HRP*.
When some of the variations of this enzyme are used as labels for antibodies, they go by names such as *anti-mouse*, *anti-rabbit*, and worryingly...*anti-human*.

MAP-kinase-kinase-kinase

Mitogen-activated protein (*MAP*) *kinases* are proteins that respond to extracellular stimuli (mitogens, *e.g.* a chemical or protein) and which regulate various cellular activities, such as gene expression, mitosis, differentiation, and cell survival or death. As you might expect, a protein that acts upon a *MAP-kinase* is called *MAP-kinase-kinase.*

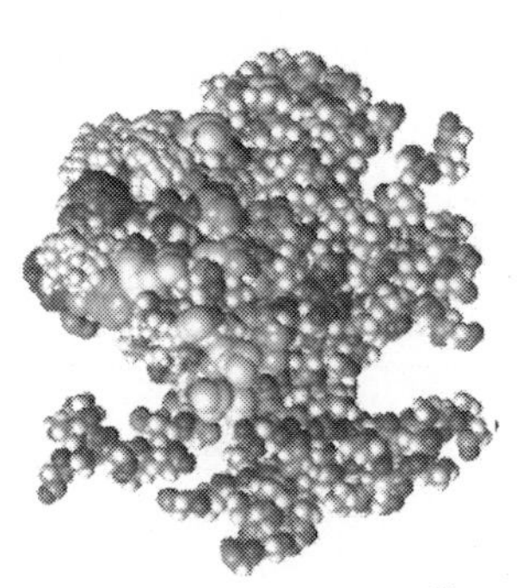

Structure of *MAP-KKK*[10].

This can go up another level, until we have the wonderfully-named *MAP-kinase-kinase-kinase* - a protein which acts on *MAP-kinase-kinase,* which acts on *MAP-kinase,* as a part of intracellular signaling [11]. Now, if yet another protein were to act on *that* molecule, I wonder what they'd call it...hmmm. To get around the long name, this protein is often shortened to *MAP3K,* or more worryingly *MAP-KKK.*

Old Yellow Enzyme

This is a *flavoprotein* that reversibly oxidises *NADPH* to *NADP* and a reduced acceptor. In fact, *yellow* enzymes are any of a number of enzymes having a *flavin* as a prosthetic group [12]. Historically, *NADPH dehydrogenase* (occurring in plants and yeast) was called the *Old Yellow Enzyme* to distinguish it from *D-amino acid oxidase,* known as, of course, the *New Yellow Enzyme.*

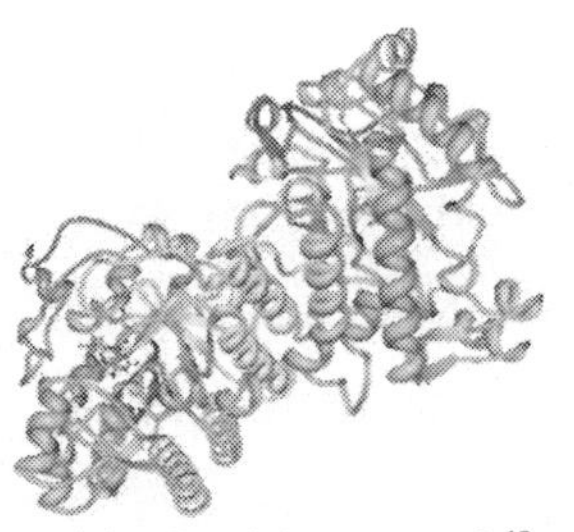

Old yellow (the enzyme) [13], not *Old Yeller* (the movie[14])

Nagarse

Nagarse is also known as *subtilisin*. It seems to be used to break apart proteins and DNA strands for analysis purposes. Its source is a bacterium and a leech (*hirudo medicinalis*). *Nagarse* is actually a trade name, named after the Nagarse (sometimes written 'Nagase') Company of Osaka, Japan.

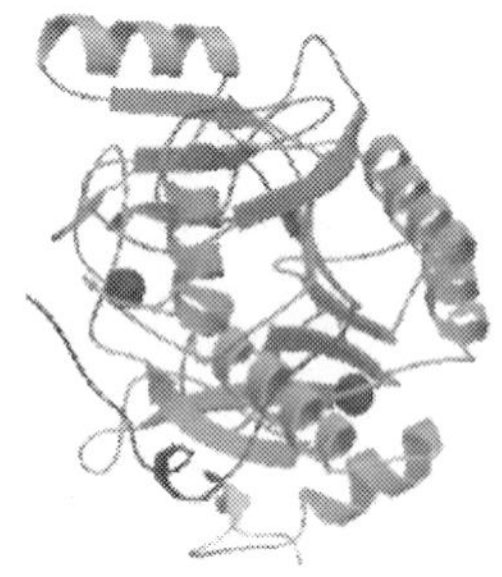

Structure of *nagarse* [15].

Miraculin

The red berries of the miracle fruit[16], with below, the listing of the amino acid residues of *miraculin* [17].

Miraculin is a *glycoprotein* extracted from the West African miracle fruit shrub [18]. *Miraculin* itself is not sweet, but, once exposed to *miraculin*, human tastebuds perceive ordinarily sour foods, such as citrus, as sweet for up to an hour afterwards. Because the miracle fruit itself has no distinct taste, this taste-modifying function of the fruit had been regarded as a miracle - hence the name of the shrub.

Sex Muscle Abnormal Protein 5

Structure of *SMAP5*[19]

The European Bioinformatics Institute houses the Macromolecular Structural Database, and for quite a few years the most downloaded structure from that site was PDB code 2SEM: it contained a protein called *Sex muscle abnormal protein 5*. Obviously, search engines had indexed all the PDB files, and you can guess what kind of searches had returned this structure!

The protein comes from *Caenorhabditis Elegans*[20], which is a tiny nematode worm (about 1 mm long) and was the first true animal to have its genome completely sequenced. As the amount of dubious material available on the internet has grown, interest in this protein from persons 'outside the scientific community' has declined. On a similar theme, there's another molecule called *SexA I*, which is an enzyme from *Streptomyces exfoliatus* that cleaves DNA.

Gonadoliberin

This name sounds like this protein makes a guy's most valued possessions drop off. In fact it's a hormone that is involved in controlling the reproductive cycle of many creatures, including humans.

Splendipherin

This splendidly named protein is a sex attractant pheromone used by the Australian 'magnificent tree frog' (*Litoria Splendida*), and was the first pheromone ever to be found in frogs [21]. It comprises 25 amino acid residues, and is water soluble. It's exuded by male frogs who use it to attract females. Splendid!

Magnificent Tree Frog, and the amino acid sequence of its splendiferous pheromone [22].

DNA origami

In this 'unfolding' new development, strands of DNA can be 'stitched' together to make nanoshapes [23]. The picture (right) shows AFM images of just some of these shapes, including nanostars, and nanosmileys. You can even write nanomessages using nanoletters, or draw nanomaps of the world (see picture below).

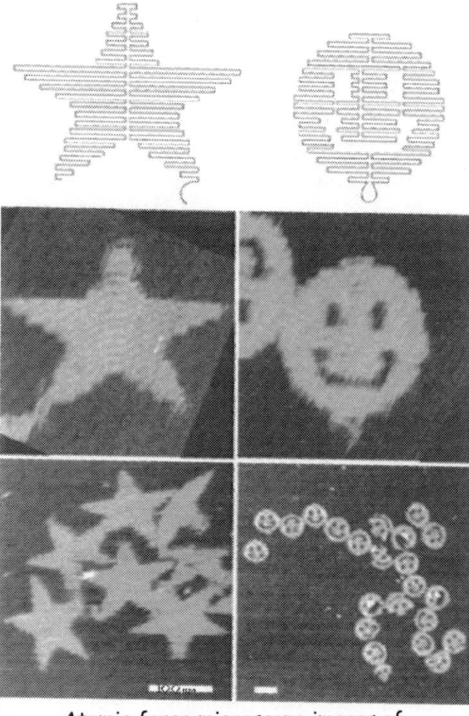

Atomic force microscope images of nanostars, nanosmileys, messages and even a map of the nanoworld. (scalebar = 100 nm). The drawings at the top show how the strand of DNA was laid down to form these shapes.
(Reprinted by permission of Macmillan Publishers Ltd, *Nature* Publishing Group, from ref.[23]).

Other protein names

There are now online databases [24,25] which contain the names of thousands of proteins, some of which are very strange indeed, including: *relaxin, survivin, fidgetin, dynamin, fukutin, antiquitin, mortalin, prohibitin, herculin, giantin, orphanin, semaphorin, arrestin, defensin, recoverin, cabin, rabin, rasputin, aladin, scramblase, Bad, Boo, CARDIAK, Casper, CASH, CLAP, DEDD, Flame, MADD, SODD, TANK, TRAMP, TRANCE, TWEAK, convertin, preconvertin, accelerin, thrombomodulin, ubiquitin, stuart* and *christmas.*

The names are often descriptive of the function of the protein or its effect upon its host (usually a lab mouse or fruitfly). For example, *stargazin* derives from mutant epileptic mice that just lie on their backs and stare at the stars.

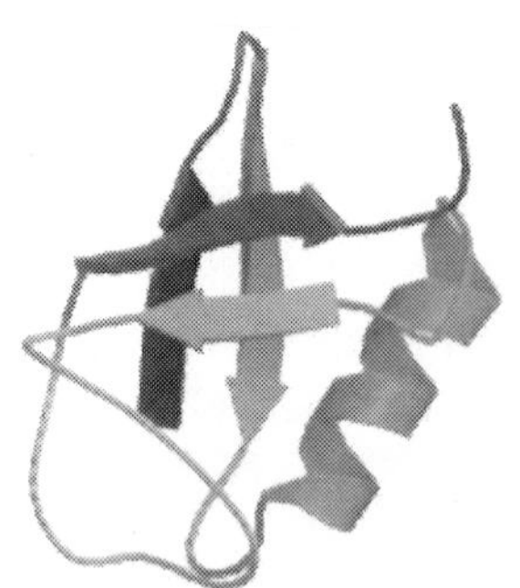

The structure of the ubiquitous protein, *ubiquitin* [26].

And there's a protein, which when absent from the rockcress plant, results in an altered leaf morphology. For this reason (and probably also for a laugh) its discoverers termed it *knobhead.*

Another example is called *dam methylase,* which is part of the *dam* modification gene of *E. coli,* and does a *dam* good job !

And there's whole series of yeast proteins called *SIR.* They are known as *Sir Silencer* molecules because they are involved in silencing and repression, although they sound more like one of King Arthur's Knights of the Round Table. Perhaps *Sir Silencer* might be gallant enough to rescue the *damsyl* in distress we met earlier...

References

[1] Image created from the 3D structural file obtained from http://www.pdb.org; structural file PDB ID:1ba3, N.P. Franks, A. Jenkins, E. Conti, W.R. Lieb, P. Brick, "Structural basis for the inhibition of firefly luciferase by a general anesthetic", *Biophys. J.* **75** (1998) 2205.

[2] R. Apitz-Castro, S. Béguin, A. Tablante, F. Bartoli, J.C. Holt, H.C. Hemker, *Thromb. Haemost.* **73** (1995) 94.

[3] Protein sequence at : http://ca.expasy.org/cgi-bin/niceprot.pl?Q9W747

[4] D.T. Jones, R.R. Reed, *Science* **244** (1989) 790.

[5] Image used is in the domain obtained from http://www.pdb.org; structural file PDB ID:1GG2, M.A. Wall, D.E. Coleman, E. Lee, J.A. Iniguez-Lluhi, B.A. Posner, A.G. Gilman, S.R. Sprang, S.R., "The structure of the G protein heterotrimer Gi alpha-1 beta-1 gamma-2" *Cell* **83** (1995) 1047.

[6] L.G. Tilney, *J. Cell Biol.* **69** (1976) 73.

[7] Image created from public domain pdb file obtained from http://www.pdb.org; structural file PDB ID:1bun, P.D. Kwong, N.Q. McDonald, P.B. Sigler, W.A. Hendrickson, "Structure of beta 2-bungarotoxin: potassium channel binding by Kunitz modules and targeted phospholipase action" *Structure* **3** (1995) 1109.

[8] O. Shakhman, M. Herkert, C. Rose, A. Humeny, C.-M. Becker, *J. Neurochem.* **87** (2003) 598.

[9] Image used is in the public domain obtained from http://www.pdb.org; structural file PDB ID:2nz2, T. Karlberg, *et al.*, "Crystal structure of human argininosuccinate synthase in complex with aspartate and citrulline".

[10] Image used was created from the pdb data file obtained from http://www.pdb.org; structural file PDB ID:2cu1, K. Inoue, T. Nagashima, F. Hayashi, S. Yokoyama, "Solution structure of the PB1 domain of human protein kinase MEKK2b", to be published.

[11] K. Yamaguchi, K. Shirakabe, H. Shibuya, K. Irie, I. Oishi, N. Ueno, T. Taniguchi, E. Nishida, K. Matsumoto, *Science* **270** (1995) 2008.

[12] R.E. Williams, N.C. Bruce, *Microbiol.* **148** (2002) 1607.

[13] Image used is in the public domain obtained from data at http://www.pdb.org; structural file PDB ID:1oyb, K.M. Fox, P.A. Karplus, "Old yellow enzyme at 2 A resolution: overall structure, ligand binding, and comparison with related flavoproteins", *Structure*, **2** (1994) 1089.

[14] http://en.wikipedia.org/wiki/Old_Yeller_(1957_film)

[15] Image used is in the public domain obtained from http://www.pdb.org; structural file PDB ID:1cse, W. Bode, E. Papamokos, D. Musil, "The high-resolution X-ray crystal structure of the complex formed between subtilisin Carlsberg and eglin c". *Eur. J. Biochem.* **166** (1987) 673

[16] http://commons.wikimedia.org/wiki/Image:Miracle.jpg, public domain.

[17] S. Theerasilp, H. Hitotsuya, S. Nakajo, K. Nakaja, Y. Nakamura, *J. Biol. Chem.* **264** (1989) 6655.

[18] K. Kurihara, L.M. Beidler, *Science* **161** (1968) 1241.

[19] Image used is in the public domain obtained from http://www.pdb.org; structural file PDB ID:2sem, J.T. Nguyen, C.W. Turck, F.E. Cohen, R.N. Zuckermann, W.A. Lim, "Exploiting the basis of proline recognition by SH3 and WW domains: design of N-substituted inhibitors", *Science* **282** (1998) 2088.

[20] S.G. Clark, M.J. Stern, H.R. Horvitz, *Nature* **356** (1992) 340.

[21] P.A. Wabnitz, J.H. Bowie, J.C. Wallace, M.J. Tyler, B.P. Smith, *Nature* 401 (1999) 444.

[22] http://commons.wikimedia.org/wiki/Image:Litoria_splendida.jpg, photographer: liquidGhoul, GNU Free Document license.

[23] P.W. Rothemund, *Nature* **440** (2006) 297.

[24] Entrez Gene: http://www.ncbi.nlm.nih.gov/sites/entrez

[25] Online Mendelian Inheritance in Man: http://www.ncbi.nlm.nih.gov/sites/entrez?db=omim

[26] Image in public domain, taken from www.pdb.org, pdb id: 1ubq, S. Vijay-Kumar, C.E. Bugg, W.J. Cook, Structure of ubiquitin refined at 1.8A resolution, *J. Mol. Biol.* **194** (1987) 531.

4. Genes with Silly or Unusual Names

Genetics has a rich history of 'creative' nomenclature. When scientists discover a new gene, they often name it based upon its function, or the appearance of the organism that has or doesn't have the gene [1,2,3,4,5]. For example, a gene responsible for the development of a fly's eyes is called *eyeless*, since files without this gene have no eyes. With names such as this one, it's easy to work out how the gene got its name. But other genes names are not so obvious. Many of the best gene names come from research into fruitflies (see photo), since scientists can play around with their genome to their heart's content and make as many weird mutants as they like, without anyone complaining. There are hundreds of strange and wonderful names, with some of the best listed below.

Fruitfly (*Drosophila*) (Photo: Sanjay Acharya [6])

However, some gene names have recently become somewhat controversial, as the analogues of genes that were originally found and named in flies are now being found in humans, and some are linked to diseases. It's quite funny when a gene is called *turnip* because it makes flies very stupid, but when the same gene is found in humans it's a different story. Imagine a doctor telling a mother that her child will have learning difficulties because he has inherited the *turnip* gene! Or worse, the *one-eyed pinhead* gene! As a result, many of these gloriously named genes may have to be renamed in the future - but until then, we can revel in their inventiveness.

Gene names

agoraphobic, reticent, shy

Fly larvae with these genes look normal, but never crawl out of the egg, as if they were afraid of the big wide world.

agnostic

Mutant flies cannot decide between odours, just as human agnostics can't make up their minds.

always early, british rail

In *always early*, only the early stages of sperm cell production are seen, which makes the male flies infertile. This gene is suppressed by the *british rail* gene (UK trains are renown for being late)!

amontillado

Causes the larvae not to hatch. Its name derives from Edgar Allan Poe's story "The Cask of Amontillado", in which a man is walled-in while still alive. Amontillado is also a type of sherry.

Left: A glass of amontillado sherry, with olives [7].

antikevordian

Named after the (in)famous Dr Kevordian who helps people commit suicide in the USA. This gene prevents the programmed death of certain plant cells.

archipelago

This affects the distribution of cells in the eyes, so there are islands of one type cell in a 'sea' of another type.

ariadne

This gene allows the growing nerve cells to find their targets, just as Ariadne helped the mythological Greek hero Theseus to find his way in Minotaur's labyrinth by giving him a ball of string.

arylsulfatase E

Although the gene name itself is not that special, it has an unfortunate shorthand symbol *(ARSE).*

asteroid

This has similar characteristics to another gene called *star,* and its name means 'starlike'.

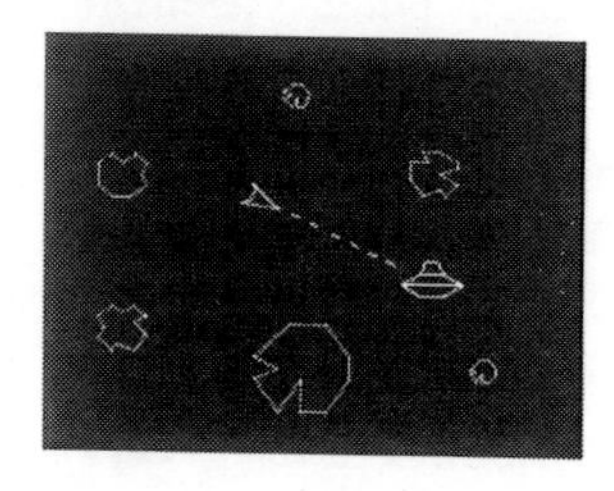

Left: Simulation of a screen shot of the computer game 'Asteroids'.

aurora, borealis

aurora is involved in creating the 'poles' of the spindle during cell division, its absence leads to formation of only one pole and a characteristic circular arrangement of chromosomes. Thus, it is like the circular, wavy auroras seen as the Northern or Southern lights at the Earth's Poles. As you might expect, the *borealis* gene is associated with it.

baboon

Named since mutant fly larvae have enlarged anal pads, like those seen on baboons [8].

Left: A real baboon
(Photo: Dick Mudde [9]).

barentsz, scott of the antarctic, shackleton

During cell division (mitosis), the cells wrap around to form a spindle shape, with the two ends being called 'poles'. Genes which affect the development of these poles, or which stop certain structures getting to them, are often named after famous Polar explorers who likewise failed to get to their North or South Poles. These include Robert Scott, Sir Ernest Shackleton and William Barentsz.

beaten path, sidestep

The first gene codes for a protein which is required for nerve cells to separate from a bundle and find their own path into a muscle. A related gene is called *sidestep*.

big brain, brainiac, minibrain

With the first two genes, mutants appear to have malformed large brains. With *minibrain*, the brain is smaller.

blistery

Causes blisters on the wings of flies. This prevents them flying. Other genes which have the same effect have been named after flightless birds, such as *moa*, *kakapo*, *rhea*, and *piopio*.

'Boat genes': ***kayak*, *punt*, *canoe*, *coracle***

The 'boat' class of fly mutants all display a prominent hole in the embryo's back, which make it resemble a boat when seen in the microscope.

botch*, *pratfall*, *misstep*, *slipshod

These cause irregular segmentation in the fly embryo [10].

bruchpilot

Maybe they had a pilot named 'Quax'.
(Photo: US. Govt. Military Air Force Maxwell [11]).

This means 'crash pilot' in German, and comes from a 1930s German movie called "Quax, der Bruchpilot". This was about a pilot who kept on crashing his plane, but always survived. Flies with this mutant gene do the same - they cannot fly properly and continually crash [12].

bag of marbles

Female mutants have ovaries with numerous tumours and partly formed cells - hence the name [13].

breathless

Involved in development of the windpipe.

buttonhead

Regulates segmentation in a fly's head. Mutants have a malformation of the pharynx resembling a button stuck to the head.

callipyge

Callipyge is Greek for 'beautiful buttocks', and this gene is named after the goddess Aphrodite Kallipygos, who exhibited those characteristics. A mutation in this gene lets sheep convert food into muscle 30% more efficiently yielding sheep with "big and muscular bottoms".[14,15]

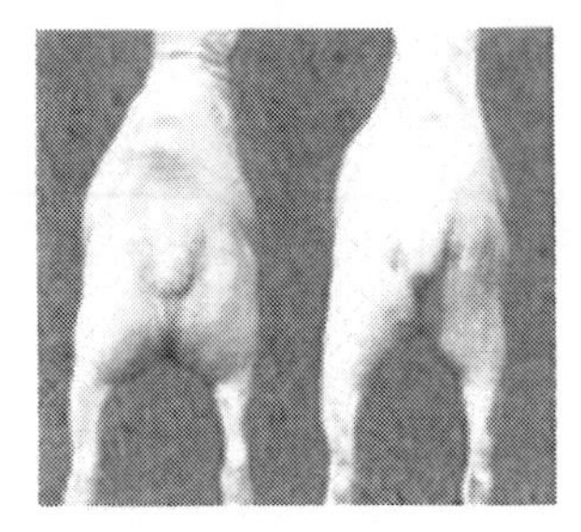

Left: I don't fancy yours much...
A *callipyge* sheep (left) shows more muscular hindquarters than a normal sheep (right).
(Photo left: used with permission of Brad A. Freking of the US Meat Animal Research Center [16], and Brad Jirtle of Duke University [17]).

cap'n'collar

This is a fruitfly gene that controls the appearance of the mandible and back of the head. It works in conjunction with the *deformed* gene which if mutated causes the head to be malformed.

castor, pollux

A pair of genes that determine the symbiotic relationship between the Japanese Lotus plant and bacteria in its roots. They were named after the twin stars in the constellation of Gemini.

celibate

Male flies are attracted to females, but never mate.

cerberus

Causes mutants with extra heads (like the 3-headed dog in Greek mythology).

charybde (or ***charybdis***), ***scylla***

These 2 genes are named after mythological monsters said to have lived in the sea between Sicily and Italy that posed a threat to the passage of ships. They were so close together that if sailors avoided one, they'd be caught by the other. The simultaneous loss of both these genes generates flies that are more susceptible to reduced oxygen concentrations, but having either one results in reduced growth.

chickadee*, *quail*, *warbler

Named after birds which have small eggs, since female flies with this gene produce smaller than normal eggs.

chiffon*, *gauze*, *rayon*, *satin*, *crepe*, *muslin

These genes control the appearance of the membrane around the fly embryo, and are named after descriptive types of material. The first four make the membrane very thin, and the last two change its appearance.

cheap date*, *amnesiac*, *lush

Mutants with the *cheap date* gene don't like alcohol! They also have a poor memory, and so the gene is also called *amnesiac*. Conversely, mutants with the *lush* gene love alcohol.

cleopatra*, *asp

A mutant with both these genes results in death, just as Queen Cleopatra of Egypt allegedly killed herself with the bite of an asp.

clockwork orange, clock, period, timeless, doubletime

These are all 'clock' genes, which tell cells the rate at which things should happen. One gene with a very pronounced effect on biological clock activity was found to code a transcriptional factor containing a basic helix-loop-helix and an ORANGE domain. It was therefore called *clockwork orange*[18], in reference to the Anthony Burgess novel and Stanley Kubrick film.

clown — This makes a fly's eyes have a characteristic red and white appearance, resembling a circus clown.

coitus interuptus — Makes mutant male flies have sex for only half the usual time.

columbus — Named after the explorer who got lost trying to find India (although he ended up discovering the New World). Likewise, mutants with this gene cause germ cells to become lost on their way to the gonad.

cookie monster — A gene involved in cell division which is named because it makes the cells look like a 'whole bunch of cookie monster eyes'[19]. For those not familiar with children's TV, the Cookie Monster is a blue Muppet with big eyes in 'Sesame Street'.

croquemort — This means 'undertaker' in French, and refers to the fact that dead cells are not cleaned up.

curl, curled, curly — Gives flies curly wings.

cyclops	Mutants have only one eye, like the creature in Greek mythology.
dachshund	Like the dog, mutant flies with this gene have small crippled legs.

Left: A short-haired dachshund [20].

daeh	This gene disturbs development of the head (*daeh* is h-e-a-d scrambled).
dally	Short for (cell) division abnormally delayed, *i.e.* it dallies.
dead ringer	Fly embryos with this gene exhibit rings around their fore and hindguts, then die.
decapentaplegic	Causes defects in 15 structures in the fly's body, *e.g.* claws, wings, legs, abdomen, *etc* [21].
dickkopf	Dick-head in German(!) is involved in determining the structure of the head [22].
dissatisfaction	Involved in many aspects of sexual behaviour!
domeless	The fly larvae are not their usual dome shape. Also, at the time of its discovery, the London Millennium Dome was just being built [23].
dreadlocks	Causes the nerve cells to become tangled up, resembling a Rastafarian hairstyle.

dribble	Causes cell proteins to move back and forth between the nucleus and the cytoplasm, like a basketball player 'dribbling' the ball [24].
droopy	Makes wings droop.
drop dead	Mutant flies' brains deteriorate, so they stagger and die after about 10 days.
ebony	Makes the fly's body colour shiny black [25].
egalitarian	All 16 egg cells develop identically [26].
eiger	Named after the mountain which has claimed so many mountaineers' lives, this gene causes cell death.

Left: The North face of the Eiger [27].

embargoed	These mutants cannot export nuclear (as in cell nucleus) material properly.
exuperantia	Named after a Christian slave who was beheaded, since the gene affects development of the head.
eye missing, eyeless, eyes absent	All affect the eyes, making them smaller or absent.
farinelli	Named after the legendary *castrato* singer Farinelli [28], this snapdragon gene produces sterile male flowers.

fear of intimacy

Individual cells normally come together to form the embryo. With this mutant, the cells stay apart, as if they were shy...

fruity, fruitless, icebox

fruity causes the male flies to be disinterested in females. This was later renamed to *fruitless* to be more politically correct. *icebox* has the same effect in female flies - they lose interest in males.

fuculokinase

This gene too has an unfortunate shorthand symbol, *FucK*. Similarly, the *E. coli K*-12 gene has proteins that have been named *Fuc-U* and *Fuc-R* [29].

furry

Makes the wings furry.

genderblind

Male flies can't tell the difference between males and females and attempt to copulate with both [30].

Genghis Khan

This is a fly gene that collects together lots of *actin* proteins, just like the powerful Mongolian emperor collected a large army and conquered most of Asia [32]. There's also a human gene found on the *Y*-chromosome that's often nicknamed the 'Genghis Khan' gene since it's believed to have come from the man himself, and was responsible for his aggressive, psychopathic behaviour. Due to his policy of killing all the captured men and personally raping all the women, he is estimated to now have 16 million descendants in Asia. This means that 1 in every 200 males in Asia is descended from Genghis Khan [33].

He's the daddy! Portrait of Genghis Khan from the National Museum, Taipei [31].

giant	Fly development is longer than normal, leading to flies nearly twice their usual size.
glass-bottomed boat	This produces transparent larvae.
gleeful	This gene is necessary for muscle development, and got its name from being related to another gene codenamed *Gli*[34].
grim, reaper, sickle	These genes cause cell death.
groucho	Just like Groucho Marx, this gene causes mutants to have an unusually large number of bristles on their face.
grounded	Flies cannot, er, fly, due to abnormal flight muscles.
grunge, teashirt	*Grunge* messes up the expression of the *teashirt* gene, just as followers of Grunge music wear dirty messy T-shirts [35].
hamlet	Named for Hamlet's "to be or not to be" soliloquy because it affects development of cells descended from the fruitfly IIB cells [36].
Harry Potter	This mice & human gene is critical in achieving puberty, hence its nickname after the fictional boy wizard. Its official name is *GPR54*.

hearsay, evander

Zebrafish with these mutant genes lack an ear and a jaw. The original name was *evander*, named after the boxer Evander Holyfield who famously had part of his ear bitten off in a fight with Mike Tyson.

heartless, brokenheart, tinman

These mutants fail to develop a proper heart. *tinman* is named after the character in the 'Wizard of Oz' who sings the song "If I only had a heart...".

hedgehog, Sonic hedgehog, Indian hedgehog, Desert hedgehog, Tiggy-Winkle hedgehog

Fruitfly embryos with these mutated genes made them look like hedgehogs, since their spines grew all over their body rather than just in specific places - so it was named the *hedgehog* gene. The vertebrate analogue was called *Sonic Hedgehog* (Shh) in order to distinguish it from its insect version [37]. Geneticists studying other organisms are happy to elaborate on this when they can get away with it, and there are other variety of hedgehog genes called *Indian hedgehog* (Ihh), *Desert hedgehog* (Dhh), and even *Tiggy-Winkle hedgehog* (TWhh)!

Left: A real hedgehog [38].

hip, hop

These genes program proteins to fold into the correct 3D structure, like breakdancers.

Homer

This is found in mammals and is responsible for development of the brain [39]. It's named after the cartoon character in need of brain development, Homer Simpson [40].

indy	Acronym for 'I'm Not Dead Yet' [41], named after the scene in the movie "Monty Python and the Holy Grail" where a guy is pushing a wheelbarrow full of corpses, shouting, "Bring out your dead!", except one of the 'corpses' is still alive, and complains, "I'm not dead yet!" As you might expect, this mutation allows the flies to live twice as long as normal.
inebriated	The mutants with this gene are uncoordinated and appear drunk.
JAK-1	A *kinase* is an enzyme that adds phosphate groups onto proteins. There are so many different *kinases* in the human body that when another one was discovered it was named *JAK*, for *just another kinase*!. The *JAK1* gene codes for this *kinase*.
jaunty	Flies with this mutant hold their wings at a strange angle, giving the appearance of being jaunty or devil-may-care.
jelly belly	Mutants have no gut muscles.
ken and barbie	Named after the dolls which have no external genitalia, neither do male and female mutants with this gene.
kenny	Mutants with this gene die if they get infected with certain bacteria. They are named after the character in the South Park cartoon series who dies at the end of every episode [42].

kiss1

This gene was originally identified by scientists in Hershey, Pennsylvania, and dubbed *Kiss1* in honour of another famous product of that town, the 'Hershey chocolate kiss' [43].

Left: A Hershey's Chocolate 'Kiss'
(Photo: PWM)

kojak

Named after the bald cop in the TV series, flies with this gene lack hair on their wings. It's also known as *shavenoid*.

kurtz

Named after the character in Joseph Conrad's book "Heart of Darkness" who, after failing to become desensitized to the horror of the Belgian Congo became insane and ultimately died. Kurtz died leaving the third station with a 'heart of darkness'. Likewise, many *kurtz* mutants die leaving part of their circulatory systems filled with black, melanotic tumors [44].

kuzbanian

Named after the Koozbanians from the original Muppet show, but with a change of spelling for copyright reasons. They were aliens with uncontrollable hair growth, and likewise, this gene causes uncontrollable growth of bristles on fly wings.

lava lamp

Causes structures within a cell to move like oil droplets in a lava-lamp.

lazarillo	This grasshopper gene acts as a guide for growing nerve cells, just as the character Lazarillo did for blind people in the 16th century Spanish novel "The Life of Lazarillo of Tormes, his fortunes and misfortunes as told by himself" [45].
lilliputian	Produces small mutants, like the characters in "Gulliver's Travels".
limo	Is involved in protein transport.
long island expressway	Named for the very long highway in New York, this gene causes the sperm cells of mutant flies to be lengthened.
lost in space	The axons of nerve cells project outwards abnormally, so they are literally lost in space... The name also refers to the classic 1960's sci-fi programme.

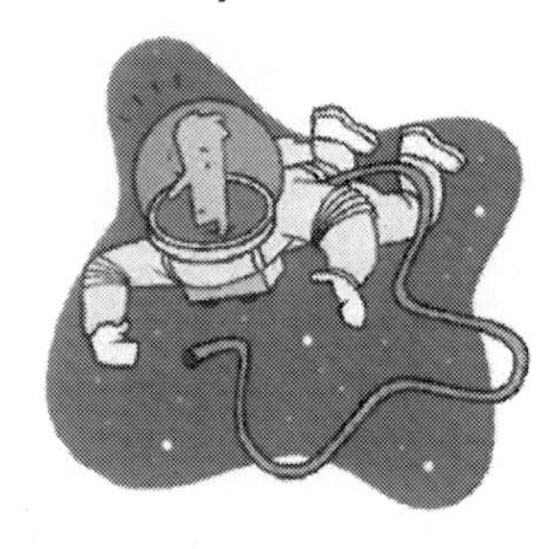

lot	Named after the Biblical character whose wife was turned to a pillar of salt, this gene makes mutants crave salt.
lunatic fringe, manic fringe, radical fringe	These genes control the patterns (fringes) on fruitfly wings[46]. A related gene is called *fringe connection.*

mad, max These genes code for proteins that bind to each other and stop bad things happening, just like Mel Gibson's character in the movie 'Mad Max'.

malvolio Named after the Shakespeare character in 'Twelfth Night' who "taste[d] with distempered appetite", this gene is required for normal taste function.

methuselah Increases lifespan (named after the long-lived Biblical character).

moleskin Prevents skin blistering.

mothers against decapentaplegic (mad), daughters against decapentaplegic (dad) These genes repress changes in the *decapentaplegic* gene (see above). The phrase "Mothers against..." was used since mothers (in the USA) often form organizations opposing various issues *e.g.* 'Mothers Against Drunk Driving'. *mad* is when mutations in the mother affect the embryo, and *dad* is when its passed on to the daughter cell [47].

moron

The convention is that functional chunks of nucleic acid usually have names ending in '-on'. For example, the coding group of 3 DNA or RNA bases is a *codon*, while a group of related genes in bacteria that are translated at once is the *operon*. It was found that the genome of one special phage (a virus that infects bacteria) had odd insertions that made them have more DNA in certain places than the other phages in the related group. Since these insertions seemed to be somewhat random but caused the phage to have *more* DNA than its cousins, the insertion was named the *moron* gene [48]. There is also a *reverse moron* gene.

noggin

Gives mutant frogs two heads ('noggin' is UK slang for head).

noose

This gene fatally interacts with another called *string* [49].

numb

Mutants lack many sensory neurons.

oskar, maggie

oskar is named after the boy Oskar Matzerath who wouldn't grow in the book "The Tin Drum" by Günter Grass. Embryos also don't grow up - they remain small. Similarly, the *maggie* gene which has the same effect is named after Maggie from 'The Simpsons' who never seems to get any older.

out at first, stranded at second

Named after the baseball terms, where the batter is out before he reaches first- or second-base. These gene causes larvae to die before they complete their first or second stages of development [50].

out cold

Low temperatures paralyse flies with this gene.

parched

The flies drink much more water than normal, and most die of water loss within 12 hours.

pavarotti

Flies with this mutation have abnormally large cells in their embryonic nervous system. It's named after the abnormally large opera singer Luciano Pavarotti [51].

Left: Pavarotti in France in 2002 [52].

Pavlov's dogs genes: ***valiet, tungus, barbos, laska, ikar, rogdi, pastrel, toi, rafael, milkah, avgust, nord, murashka, beluy, mampus, novichok, mirta, rijiy, zloday, diana, zolotistyuy, rosa, gryzun, chyorny, moladietz, dikar, drujok, jack, martik, premjera, visgun.***

These were the names of the dogs that Ivan Pavlov's used in his famous experiments on conditioned responses for which he got the Nobel prize for medicine in 1904. He discovered that if he made a signal (such as ringing a bell) whenever the dogs were fed, the dogs would salivate, even before the food arrived, showing that they'd learnt to associate the bell ring with approaching food.
As you might expect, all of the 'Pavlov's dogs' genes create fly mutants with long-term memory problems.

peanut

Cells form two nuclei, just like a peanut with two kernels.

pimples

Embryos have patches on their skins, like pimples.

pokemon

Stands for POK Erythroid Myeloid ONtogenic factor, and is responsible for the evolution of cancer. Unfortunately, the Nintendo subsidiary Pokémon USA, fearing the bad press that might result from sharing a name with a cancer-causing gene, threatened legal action in December 2005. So the gene has now been renamed *Zbtb7.*

pray for elves

Named by Suzan Lewis (one of the workers on the flybase annotation project) who was so snowed-under with work that she prayed for some 'magic little elves' to appear who would relieve her workload.

prospero Named after Shakespeare's character in "The Tempest" who could influence the fate of other characters. This gene can also influence the fate of cells.

ring The function of this gene was not originally known, but it was a 'really interesting new gene'. A domain in the protein was later called *ring finger*.

rotated penis Causes the external genitalia of flies to be rotated by about 180°.

sarah Named after another Biblical character, Abraham's wife, who was unable to have a child for many years. This gene makes mutant flies practically sterile.

shibire Japanese for 'paralysed', since it causes the flies to stop moving at certain temperatures. Since the gene-code for *shibire* is *shi*, and the temperature sensitive version is denoted with the suffix 'ts', many papers state that flies with the *shi*ts don't move a lot.

seven up, sevenless, seven in absentia

The compound eye of a fruitfly is comprised of ~750 facets each containing 8 different photoreceptor cells. Mutant flies with the *seven up* gene mis-specify the receptors in their eyes, so that the signals from R1, R3, R4, and R6 are recorded as being from R7. Flies with the *sevenless* and *seven in absentia* genes are missing the R7 receptor. Related genes are called *bride of sevenless, son of sevenless*, and *daughter of sevenless.*

Left: Four tins of 7-up (does that equal 28-up?)
(Photo: PWM)

seizure

Adult flies have a fit then become paralysed at high temperatures.

shaker, ether a go-go

Flies' legs shake uncontrollably on exposure to ether.

single minded

Flies with mutations in this gene possess a single bundle of nerve cells in their nervous systems instead of two.

slowpoke

Flies slow down and become lethargic at high temperatures.

smaug, nanos

Smaug was the name of the dragon in Tolkien's book 'The Hobbit', who drove the dwarves away from their cave. Likewise, the *smaug* gene represses the activity of the *nanos* gene (which is Greek for 'dwarf').

snafu

'Snafu' is an American military acronym for 'Situation Normal All F**ed Up'. In fruitflies with this mutated gene, the embryos are initially normal but become progressively more abnormal as they develop.

spaghetti squash

This gene causes accumulation of giant cells with bundles of chromosomes per nucleus, which resemble a tangled bowl of spaghetti.

stuck

This defect causes male flies to become stuck in the females after copulation because their appendages are held in aberrantly protruding positions. They should be so lucky...

sunday driver, redtape, roadblock, gridlock	These genes all mess up intracellular traffic and cause transport problems in cells!

Left: Gridlock in Zagreb [53].

superman, clark kent, kryptonite	Mutation of this gene produces extra stamens (male genitals) in the flowers of the plant *Arabidopsis thaliana*. A later gene was named *clark kent* as its mutation was like *superman*, only more wimpish. Later it was found that *clark kent* was not a separate gene at all, but just another form of *superman*! As you might expect, another gene called *kryptonite* suppresses the action of *superman*.
swallow	Embryos have abnormal head structures and cannot eat.
swiss cheese	The brains of flies with this gene have large holes, resembling Swiss cheese.

Left: Swiss cheese (or a fly's brain) [54].

taxi Flies have abnormal wings held stiffly outwards. They cannot fly, only 'taxi' along the ground.

technical knockout The slightest mechanical shock causes mutant flies to fall over paralysed.

thisbe, pyramus Named after the two 'heartbroken' lovers in the Roman legend of Thisbe and Pyramus, since the genes are closely linked and embryo flies with these genes have no heart [55].

thor Named after the Nordic God of Thunder who defended the Vikings, this gene protects fruit flies from disease.

tigger, gypsy, hobo, Jordan element *tigger* is named after the bouncy character in the book 'Winnie the Pooh'. This gene can jump to different locations in the human genome. The next two genes are similarly named for their wandering nature, while *Jordan element* is named after the basketball player Michael Jordan for its extraordinary capacity to jump around the *volvox* genome.

trailer hitch Fly eggs have a fused dorsal appendage which looks like a trailer hitch in a car [56].

tribbles	Causes cells to divide uncontrollably, just like the furry creatures from Star Trek that reproduced as fast as they ate.
tudor, staufen, vasa	These genes cause the mutants to have no offspring, just like various Royal families in history, e.g. the Royal Tudor family of England in the 16th century, or the Vasa Royal family from Sweden & Poland. *staufen* is particularly appropriate, since as well as having no offspring, the last Staufen (from the German dynasty Hohenstaufen) was beheaded, which is analogous to the head defects seen with this mutant gene.

Left: King Henry VIII (a Tudor) [57].

turnip, dunce, cabbage, rutabaga, radish	Flies with these genes exhibit learning difficulties, and are as stupid as vegetables!
twit	The shorthand symbol for the *twilight* gene, involved in cell reproduction.
van gogh, starry night	these produce swirling patterns in the wing hairs, which resemble the 1889 painting "The Starry Night" by artist Vincent van Gogh.
vulcan	Vulcan was the Roman god of fire who was thrown down from heaven by his father Jupiter, and broke both his legs. This gene creates mutants with malformed legs.

wee1 — This yeast gene makes mutant cells divide before they are ready, so they are much smaller than normal. It was named by a Scottish scientist, hence *wee ones*).

werewolf — Plants with this mutant gene have very hairy roots.

'Wine genes': ***chablis***, ***frascati***, ***merlot***, ***retsina***, ***riesling***, ***cabernet***, ***grenache***, ***chardonnay***, ***chianti***, ***pinotage***, ***sauternes***, ***weissherbst***, ***zinfandel***, ***freixenet***, ***yquem*** — These are all zebrafish genes named after red and white wines, which correspond to whether the embryos contained blood or not [58]. Related to these is the *moonshine* gene which was named for its sparkle as a result of rays of light reflecting off the shiny embryo.

wishful thinking — Named after the discoverers' belief that they'd eventually find a link between it and the proper development of the nervous system [59].

wunan — Germ cells with this mutant gene wander around and do not complete their migration successfully. This is similar to the character Wunan in the Chinese legend, who in pig form wandered about stealing food and not doing what he was supposed to.

yippee

Named after a graduate student's reaction when she finally managed to clone this gene. She had the habit of writing 'yippee' in the margins of her notebook when an experiment went right, and this became the name of the gene.

yuri

Named after the world's first astronaut, Yuri Gagarin, because mutant flies with this gene have problems with gravity.

Left: Yuri Gagarin [60].

ZapA

A bacterial gene named after musician Frank Zappa [61].

zero population growth

Males and females are both infertile.

18-wheeler

Mutant flies produce larvae with 18 stripes.

A zebrafish [62]

Zebrafish ear genes [63]:

big ears,
little ears,
dog eared,
headphones,
boxed ears,
backstroke,
what's up?,
van gogh,
earache,
spock,
ear plugs,
half stoned,
rolling stones,
einstein,
keinstein,
stein und bein,
menhir,
mind bomb,
starmaker,
mariner,
orbiter,
sputnik,
gemini

Zebrafish have transparent embryos, so it's easy to study their development. A lot of work has been done looking at the genes which control the development of the ear. Genes which produce malformed ear structures are given names like *dog eared or boxed ears.* The *spock* mutation produces pointy ears (like Mr Spock of Star Trek), while *headphones* creates greatly extended ears, and *van gogh* (a different gene to the one of the same name found in fruitflies), *earache, earplugs* and *what's up?* produce deafness!

The inner ears contain small stonelike structures which help with hearing and balance. Genes which affect the development of these often have 'stone' in their name, such as *half stoned* (only half the stones are present), *rolling stones* (the stones are found in unusual places), *stein und bein* (German: stone and bone), *menhir* (a large standing stone like those in Stonehenge), *einstein* ('one stone' in German) and *keinstein* (= kein einstein, 'not one stone'). Some mutations cause the stones to change shape from round to star-like, so we have genes named after spacey things, such as: *starmaker, mariner, orbiter, sputnik* and *gemini.*

Miscellaneous others

There are literally thousands of known genes, and a remarkable number of them have peculiar names. Below is just a personal selection of my favourites - see if you can guess how they got their names...

abrupt, anachronism, ATM, bang senseless, bang sensitive, bagpipe, bazooka, bientot (French: soon), *bifocal, blown fuse, bowel* (*brother of odd with entrails limited*), *brain tumor, brakeless, bric à brac, buffy, bumper-to-bumper, cacophony, cannonball, cappuccino, capricious, clootie dumpling, clueless, cockeye, comatose, concertina, couch potato, cubitus interruptus, currant bun, daughter killer, DDT, deadpan, derailed, disco-related, dog of glass, double parked, doublesex, double-time, dumbfounded, escargot, faint sausage, floating head, frizbee, GAGA, gammy legs, gluon, gooseberry, grainyhead, gutfeeling, half pint, headcase, helter-skelter, he's not interested, highwire, hindsight, hoi-polloi, hunchback, jetlag, just odd knobs, king tubby, kismet, knirps, kon-tiki, ladybird early, ladybird late, lame duck, lemming, lesbian, liquid facets, males absent on the first, maelstrom, mastermind, members only, merlin, Mom* (more mesoderm), *myoblast city, nervous wreck, neurotic, nuclear fallout, one-eyed pinhead, pacman, pickpocket, pipsqueak, Pop* (posterior pharynx defect), *presto, quick-to-court, radar, rags, rubbish, saxophone, screw, scribbled, scribbler, sex combs reduced, sex lethal, shotgun, shuttle craft, singles bar, skittles, slamdance, slingshot, sloppy pair 1 and 2, Small Mothers Against Death, smurfl, sort of bloodless, spinster, splat, split ends, spotted dick, strawberry notch, sticks and stones, stumps, takeout, terribly reduced optic lobes, thousand points of light, tramtrack, vibrator, windbeutel* (German: "cream puff" or "windbag"), *yorkie, you too, zipper.*

References

[1] M.R Seringhaus, P.D Cayting, M.B Gerstein, *Genome Biology* **9** (2008) 401.
[2] M. Vacek, *Am. Sci.* November/December 2001,
http://www.americanscientist.org/template/AssetDetail/assetid/14672
[3] http://www.flynome.com
[4] http://flybase.bio.indiana.edu/
[5] http://tinman.vetmed.helsinki.fi/eng/intro.html
[6] Wikimedia Commons, GNU free document licence; http://commons.wikimedia.org/wiki/Image:Fruit_fly5.jpg
[7] http://commons.wikimedia.org/wiki/Image:Del_Duque_Amontillado_Sherry.jpg, photo: Hashashin, GNU Free Documentation License.
[8] T. Brummel, S. Abdollah, T.E. Haerry, M.J. Shimell, J. Merriam, L. Raftery, J.L. Wrana, M.B. O'Connor, *Genes & Dev.*, **13** (1999) 98.
[9] Wikimedia Commons, public domain; http://commons.wikimedia.org/wiki/Image:Baviaan1.JPG
[10] T. Schüpbach, E. Wieschaus, *Genetics* **129** (1991). 1119.
[11] Wikimedia Commons, public domain; http://commons.wikimedia.org/wiki/Image:B-24_Kopfstand.jpg
[12] D.A. Wagh, T.M. Rasse, E. Asan, A. Hofbauer, I. Schwenkert, H. Dürrbeck, S. Buchner, M.-C. Dabauvalle, M. Schmidt, G. Qin, C. Wichmann, R. Kittel, S.J. Sigrist, E. Buchner, *Neuron*, **49** (2006) 833.
[13] D.M. McKearin, A.C. Spradling, *Genes & Dev.*, **4** (1990) 2242.
[14] N.E. Cockett,.S.P. Jackson, T.D. Shay, D. Nielsen, S.S. Moore, M.R. Steele, W. Barendse, R.D. Green, M. Georges. *Proc. Natl. Acad. Sci. USA*, **91** (1994) 3019.
[15] B.A. Freking, S.K. Murphy, A.A. Wylie, S.J. Rhodes, J.W. Keele, K.A. Leymaster, R.L. Jirtle, T.P. Smith,. *Genome Res.* **12** (2002) 1496.
[16] http://www.usmarc.usda.gov
[17] http://www.geneimprint.com/lab/
[18] C. Lim, B.Y.Chung, J.L. Pitman,J.J. McGill, S. Pradhan, J. Lee, K.P. Keegan, J. Choe, R. Allada, *Current Biol.* **17**, (2007) 1082.
[19] J. Jiang, H. White-Cooper, *Development* **130** (2003) 563.
[20] Modified from the photo on Wikimedia Commons, photo: Igor Bredikhin, Creative Commons license, http://commons.wikimedia.org/wiki/Image:Short-haired-Dachshund.jpg; original at: http://www.dachshund-land.ru
[21] F.A. Spencer, F.M. Hoffmann, W.M. Gelbart, *Cell* **28** (1982) 451.
[22] http://en.wikipedia.org/wiki/DKK1
[23] S. Brown, N. Hu, H.J.E.Castelli-Gair, *Curr. Biol.* **11** (2001) 1700.
[24] H.Y.E. Chan, S, Brogna, C.J. O'Kane, *Mol. Biol. Cell* **12** (2001) 1409.
[25] C.B. Bridges, T.H. Morgan, *Publs Carnegie Instn.* **327** (1923) 1.
[26] T. Schupbach, E. Wieschaus, *Genetics* **129** (1991) 1119.
[27] http://commons.wikimedia.org/wiki/Image:Eiger_2415.jpg, photo: Dirk Beyer, GNU Free Documentation License.
[28] http://en.wikipedia.org/wiki/Farinelli
[29] See the following webpages for details:
http://biocyc.org/ECOLI/new-image?type=GENE-IN-MAP&object=EG10350;
http://biocyc.org/ECOLI/new-image?type=GENE-IN-MAP&object=EG10355;
http://biocyc.org/ECOLI/new-image?type=GENE-IN-MAP&object=EG10353
[30] H. Augustin, Y. Grosjean, K. Chen, Q. Sheng, D.E. Featherstone. *J. Neurosci.* **27** (2007) 111.
[31] Wikimedia Commons, GNU free document license, http://commons.wikimedia.org/wiki/Image:Chengiz-khan.jpeg
[32] L. Luo, T. Lee, L. Tsai, G. Tang, L.Y. Jan, Y.N. Jan, *Proc. Natl. Acad. Sci. USA*, **94** (1997) 12963.
[33] T. Zerjal, Y. Xue, G. Bertorelle, *et al.*, *Am. J. Hum. Genet.*, **72** (2003) 717.
[34] E.E.M. Furlong, E.C. Andersen, B. Null, K.P. White, M.P. Scott, *Science*, **293** (2001) 1629.
[35] A. Erkner, *et al.*, *Development* **129** (2002) 1119.
[36] A.W. Moore, L.Y. Jan, Y.N. Jan, *Science* **297** (2002)1355.
[37] http://en.wikipedia.org/wiki/Sonic_hedgehog
[38] http://commons.wikimedia.org/wiki/Image:Igel.JPG, photo: Gibe, GNU Free documentation licence.
[39] P.R. Brakeman, A.A. Lanahan, R. O'Brien, K. Roche, C.A. Barnes, R.L. Huganir, P.F. Worley, *Nature*, **386** (1997) 294.
[40] Personal communication from Richard O'Brien, John Hopkins University, USA.
[41] http://whyfiles.org/shorties/070old_fly/
[42] S. Rutschmann,A.C. Jung, R. Zhou, N. Silverman, J.A. Hoffmann, D. Ferrandon, *Nature Immun.* **1** (2000) 342.
[43] H.M. Dungan, D.K Clifton, R.A. Steiner, *Endocrinology* **147** (2006), 1154.

[44] G. Roman, J. He, R.L. Davis. *Genetics* **155** (2000) 1281.
[45] http://en.wikipedia.org/wiki/Lazarillo
[46] J.Y. Wu, L. Wen, W.-J. Zhang, Y. Rao, *Science*, **273** (1996) 355.
[47] J. Sekelsk, S.J. Newfeld, L.A. Raftery, E.H. Chartoff, W.M. Gelbart, *Genetics* **139** (1995) 1347.
[48] R.J. Juhala, M.E. Ford, R.L. Duda, A. Youlton, G.F. Hatfull, R.W. Hendrix, *J. Mol. Biol.* **299** (2000) 27.
[49] H. White-Cooper, M. Carmena, C. Gonzalez, D.M. Glover, *Genetics* **144** (1996) 1097.
[50] D.E. Bergstrom, C.A. Merli, J.A. Cygan, R. Shelby, R.K. Blackman, *Genetics* **139** (1995) 1331.
[51] R.R. Adams, A.A.M. Tavares, A. Salzberg, H.J. Bellen, D.M. Glover, *Genes Dev.* **12** (1998) 1483.
[52] http://commons.wikimedia.org/wiki/Image:Luciano_Pavarotti_15.06.02_cropped.jpg, photo: Pirlouiiiit, Creative Commons ShareAlike License.
[53] http://commons.wikimedia.org/wiki/Image:Zagreb_guzva.jpg, photo: Mate Balota, public domain
[54] http://commons.wikimedia.org/wiki/Image:NCI_swiss_cheese.jpg, photo: Renee Comet, public domain.
[55] A. Stathopoulos, B. Tam, M. Ronshaugen, M. Frasch,and M. Levine, *Genes & Develop.* **18** (2004) 687.
[56] J.E. Wilhelm, M. Buszczak, S. Sayles, *Dev Cell* **9** (2005) 675.
[57] http://commons.wikimedia.org/wiki/Image:Holbein_henry8_full_length.jpg, photo of a painting by Holbein, *ca.* 1537, public domain.
[58] D.G. Ransom, *et.al. Development* **123** (1996) 311.
[59] G. Marqués, H. Bao, T.E. Haerry, M.J. Shimell, P. Duchek, B. Zhang, M.B. O'Connor, *Neuron*, **33** (2002) 529.
[60] http://commons.wikimedia.org/wiki/Image:Gagarin_space_suite.jpg, original source of photo: Russian Institute of Radionavigation and Time (www.rirt.ru), public domain.
[61] C. Wassif, D. Cheek R. Belas, *J. Bacter*, **177** (1995)5790.
[62] http://commons.wikimedia.org/wiki/Image:Zebrafisch.jpg, photo: Azul, public domain.
[63] T.T. Whitfield, M. Granato, F.J.M. van Eeden, U. Schach, M. Brand, M. Furutani-Seiki, P. Haffter, M. Hammerschmidt, C.-P. Heisenberg, Y.-J. Jiang, D.A. Kane, R.N. Kelsh, M.C. Mullins, J. Odenthal, C. Nüsslein-Volhard, *Development* **123** (1996) 241.

5. Silly Molecule Names - the Game

Another way you can enjoy this book is to turn it into a party game. You need to appoint one person to be the referee who has sole access to this book, but you can have any number of other players. To play, the referee chooses the name of a silly molecule, mineral, protein or gene and writes it on a piece of paper, and hands it to the players. Each player has to guess if the name is genuine (*i.e.* is in this book) or false (*i.e.* the referee made it up!). If the player gets the wrong answer, they have to perform a forfeit. A typical forfeit might be to drink some alcoholic beverage - although the forfeit can be anything you wish, depending upon who's playing and how daring you are... Players with the correct answer get a point each (and no forfeit), and the winner is the player with the most points at the end of the game.

The referee must think of a number of fake but believable molecule/gene names before the game starts. Here is a list of some fictitious names[1] to get you started (along with explanations in brackets in case you don't get the joke). Don't read the name aloud to the players or they might spot that they're fake too easily!:

Fake molecule names: *i-nositol* (I knows it all), *fidourine* (dog pee!), *precipitic acid*, *thiotimoline* (Isaac Asimov's famous fictitious molecule), *resurrectine*, *lazarine* (they raise the dead?), [the following are potential Viagra substitutes... *mycoxafloppin, mycoxafailin, mydixadrupin, mydixarizin, mydixadud, dixafix, i-bepokin*], *dumacium* (dumb ass), *tedium sulfate, euphorium nitrate, europium stagnate, didumdidumdium* (metal component of musical instruments?), [the following are fake names for placebos: *confabulase, gratifycin, deludium, hoaxacillin, dammitol, placebic acid, panacease, obecalp* (placebo backwards)], 2,4,6,8-*motorway* (song by Tom Robinson), *para-dice, ortho-gonal, meta-tarsal* (toe bone), *para-chute,*

---

[1] If any of these invented names turn out to actually be genuine, please let me know and I'll remove them for the next edition of this book.

A. para-tus, bi-4-nowthen, E-bi-gum (UK slang expression of surprise from Yorkshire, 'ee-bah-gum'), *meso-horneal* (me so horny), *basic acid, tertiary syphillitic acid, tri-ardourmate* (try harder, mate), *octo-pusleg, ethyl fornicate, ethyl celibate, ethyl palpitate, butnakide, pent-up-D-syre, groinograbin, Rektum reagent, humungous chloride, hymenic acid, illigitimone, Dipschitz salt, [name]-one,* (where you can replace [name] by the name of a celebrity, *e.g. brucewillisone, britneyspearsone*), *inane, nobrane, inkstane, spunkstane, hasbene, baykebene* (baked bean), *wetdreme, soddol, buggarol* (bugger all), *fonecol* (phone call), *seximone, sexigrone, dragquene, obsene, unklene, Fe-lineurine* (cat pee), *ludicrous acid, ridiculous acid.*

Fake mineral names: *fubarite* (from 'fubar'), *leverite* (leave 'er right there), *baggashite, crockashite, pylashite, heapashite, gobshite, torqinshite, dumperite, silentnite, nightlite, humpawlnite, overbite, dogfite, marmite, flyinakite, millerlite* (note: there's a genuine mineral called *millerite*), *earmite, justmite, blacknite, goodenite, notquite, leftrite, humanrite, rongunrite, website, paweyesite, hindsite, picnicsite, holdtite, lycratite, washenwite, blackenwite, witerenwite, snowite, lymphozite, leucozite* (white blood cells), *lamplite.*

Fake protein names: *tossin, murmurin, stranglin, chickin, droppin, parachutin, scratchin, hangglidin, hitchhikin, tickin, brakin, upchuckin, vomitin, nosedrippin, lynchin, paintin, decoratin, i-luvulin, hatpin, stickin, withdrawin, gonfishin, dunworkin, rintintin, dustybin, sinbin, fatfase, inadase, dogdase, teenagecrase, theatreplase, separatewase, freewase, steelygase, heat-hase, easylase, crimepase, needarase, flowervase, butkikin.*

Fake gene names: *smartallic, dogbreath, c'est la vie, here we go again, down the drain, all mouth no trousers, rock'n'roll rebel, poet's day, smashed and stoned, one-eyed trouser snake, life of brian, clinically braindead, mashed potato, multiple pile-up, one foot in the grave, ménage à trois, sh*t for brains, one-legged stalker, sex explosion, iron maiden, god save the queen, disco inferno, headbanger, unconscious air guitarist, five-legged sex beast, triple knob, catholic priest, silent budgerigar, polly wanna cracker, eyes on stalks, pussy galore, short fat hairy legs, red herring, freakazoid, stage diver, ejection seat.*